Cell Fusion

HENRY HARRIS

Fellow of Lincoln College and Professor of Pathology in the University of Oxford

Cell Fusion

THE DUNHAM LECTURES

HARVARD UNIVERSITY PRESS
CAMBRIDGE, MASSACHUSETTS
1975

Library of Congress Catalog Card No. 72–194000
International Standard Book No. 0 674 10464

PRINTED IN GREAT BRITAIN
BY W. S. COWELL LTD
AT THE BUTTER MARKET, IPSWICH

To A. F. H.

without whom experiments would not be done,
nor books written.

Preface

THIS book is based on the Dunham Lectures given at Harvard University in the autumn of 1969. The Chairman of the Dunham Lectureship Committee made it clear, when he invited me to give these lectures, that the Committee did not insist on their publication; but I have chosen to publish nevertheless, and this for two reasons. First, the receipt of many letters suggesting that I should review the subject of cell fusion leads me to hope that this book may serve some useful purpose; and second, the volume of experimental work involving the fusion of cells is growing so rapidly, that, if I do not review the subject now, I fear I never shall. A review of cell fusion may soon confront the reviewer with the same problems of selection and organization as a review of bacterial sexuality. Much of the work described in this book was done at the Sir William Dunn School of Pathology in the University of Oxford; and it was, at various times, the work of many hands. It gives me great pleasure to thank M. E. Bramwell, P. R. Cook, D. M. Grace, R. T. Johnson, O. J. Miller, M. D. C. A. Quinn, E. E. Schneeberger, G. I. Schoefl, E. Sidebottom, and J. F. Watkins for their collaboration and for their pleasant company. I am also not a little indebted to collaborators from other laboratories: Dr E. Engel of Vanderbilt University, Nashville, Tennessee; Dr N. R. Ringertz and Dr L. Bolund of the Institute for Cell Research and Genetics, Karolinska Institutet, Stockholm; and Prof. G. Klein, Dr P. Worst and Dr T. Tachibana of the Institute of Tumour Biology at the Karolinska Institutet. Many of the observations discussed in these pages would be much more obscure than they now are, if these friends from overseas had not contributed their skill and enthusiasm.

Oxford, 1969   HENRY HARRIS

Acknowledgements

I wish to thank the editor of the London *Daily Mirror* for permission to reproduce Fig. 1.1; and the editors of *Journal of Cell Science, Proceedings of The Royal Society,* and *Nature* for permission to reproduce many figures originally published by my colleagues and myself in these journals.

Contents

FIG. 1.1. One reaction to the discovery that cells from different animal species could be fused together to form viable hybrids. From the London *Daily Mirror* of 15 February 1965.

The Formation and Characteristics of Hybrid Cells

Historical

ON 13 February 1965, J. F. Watkins and I reported that an inactivated virus could be used to fuse together cells from different animal species and that the hybrid cells produced in this way were viable.[1] The newspapers of the world were not slow to appreciate the biological significance of this discovery, as Fig. 1.1, taken from the London *Daily Mirror* two days later, clearly shows. Our experiment was at once hailed as a 'break through', and thus received the ultimate accolade of scientific journalism. Science, of course, rarely, if ever, proceeds in this way; and our experiments, like all other 'break throughs', had their origins in a long history of previous work without which our own contribution would neither have been made nor envisaged. This history is not uninteresting.

Multinucleate cells in vertebrates were apparently first described by Müller,[2,3] who observed them in tumours. Robin[4] noted their presence in bone marrow, Rokitansky[5] in tuberculous tissue, and Virchow[6] in a variety of normal tissues and both inflammatory and neoplastic lesions. By the time Langhans[7,8] wrote his classical papers on multinucleate cells, an extensive literature about them already existed. The view that some of these cells were produced by fusion of mononucleate cells derived from the work of de Bary,[9] who observed that the life-cycle of certain myxomycetes involved the fusion of single cells to form multinucleated plasmodia. Lange[10] appears to have been the first to describe a process of this sort in vertebrates. Lange observed the coalescence of blood-borne amoeboid cells in the frog; and similar observations were made a little later by Cienkowski,[11] Buck,[12] and Geddes[13] in invertebrates. Metchnikoff[14] considered that the fusion of phagocytic cells to form plasmodia was a characteristic cellular defence mechansim in both vertebrates and invertebrates.

It is very probable that some of the inflammatory lesions in which multinucleate cells were observed in the last century were caused by viruses, although, of course, the viral aetiology of these conditions was not recognized until very much later. The earliest reports of multinucleate cells in lesions which can with certainty be identified as of viral origin appear to be those of Luginbühl[15] and Weigert,[16,17] who described such cells at the periphery of smallpox pustules. Unna[18] observed multinucleate cells in the skin lesions of chicken pox, and Warthin[19] observed them in the tonsils of patients with measles.

Following the introduction of tissue culture methods by Harrison,[20] numerous observations were made on cell fusion in cultures of animal tissues. The first of these appears to have been that of Lambert;[21] fifteen years later, Lewis[22] was able to list twenty-one references to observations of this kind. Warthin's observations[19] prompted Enders and Peebles[23] to examine the effects of measles virus in tissue culture. These authors found that the virus induced the cells in the tissue culture to fuse together to form multinucleated syncytia. A similar observation was made by Henle, Deinhardt, and Girardi[24] with mumps virus, by Chanock[25] with the virus of infantile croup, and by Marston[26] with a virus of the para-influenza group. Numerous other examples of this phenomenon have since been described. Okada[27,28] demonstrated that animal tumour cells in suspension could be rapidly fused together to form multinucleate giant cells by high concentrations of HVJ virus, another member of the para-influenza group.

Whether the various kinds of multinucleated cells seen in pathological lesions are capable of multiplication remains an incompletely resolved controversy to the present day; but, as early as 1916, Macklin[29] had already observed that binucleate cells in chick embryo tissue cultures could undergo mitosis, and that the mitosis sometimes gave rise to two mononucleate daughter cells. Macklin described how, in such cases, the chromosomes from both nuclei became aligned along a single equatorial plate and were then distributed by normal cell division to the two daughter cells. Macklin also observed various forms of irregular mitosis in binucleate cells, including tripolar mitosis. Fell and Hughes[30] described essentially the same process in binucleate mouse cells in culture. Here, synchronous mitosis of the two nuclei again collected all the chromosomes along a single equatorial plate, and cell division gave rise to mononucleate daughter cells with abnormally large nuclei apparently containing twice the normal number of chromosomes. By 1960, then, it was clear that cells in culture could fuse with each other spontaneously, that fusion could be induced at will by the use of certain viruses, and that binucleate cells could give rise to mononucleate daughters in which chromosomes from both the nuclei in the original binucleate cell were included in a single nucleus.

In 1960 Barski, Sorieul, and Cornefert[31] reported that when two different lines of mouse cell were grown together in culture, a new cell type eventually appeared which contained within a single nucleus the chromosomal complements of both parent cells. Barski and Belehradek[32] considered that these hybrid cells might have arisen by a process of nuclear transfer between cells and produced some cinematographic observations in support of this view. The idea that the hybrid cells might have arisen from cell fusion induced by viruses was not, at the time, favoured, although subsequent observations revealed that these cells carried SV5 virus, a potent inducer of cell fusion.[33] Sorieul and Ephrussi,[34,35] and later Ephrussi with several other collaborators,[36] extended the observations of Barski *et al.*[31] and obtained several new hybrids from mixed cultures of different mouse cell lines. Ephrussi and his collaborators carried out extensive karyological investigations on these mouse hybrid cells, and showed that they tended to lose chromosomes, but slowly and unpredictably. Littlefield,[37] by making use of parent cells with different biochemical deficiencies, devised a method for selecting the occasional hybrids that arose spontaneously in mixed cultures of these defective parents. Prior to the publication of the Harris and Watkins experiment,[1] hybridization had only been observed between different strains of mouse cells. Our contribution was threefold: (1) to show that an inactivated virus could be used to provide a general method for fusing animal cells together under controlled conditions; (2) to show that fusion could be induced between cells from widely different species; and (3) to show that the fused cells were viable. The only really surprising thing about our experiment was that it had not been done much earlier.

The mechansim of cell fusion

For our work Watkins and I chose to use the 'Sendai' virus, which has now become the standard reagent for the induction of cell fusion. That is not to say that it is necessarily the best reagent, and it is surprising that the range and efficacy of other viruses have not been extensively explored. The 'Sendai' virus is a member of the para-influenza group of myxoviruses and is very similar to, if not identical with, the HVJ virus (Haemaggluti-nating Virus of Japan). Watkins and I chose the 'Sendai' virus because Okada[27,28] had previously shown that animal tumour cells in suspension could be rapidly fused together by high concentrations of this virus, and Okada and Tadokoro[38] had shown that the ability of the virus to fuse cells together was relatively resistant to doses of ultraviolet light which destroyed its infectivity. The latter observation implied that inactivated virus could be used as the fusing agent, and the complications arising from the use of infective virus could thus be avoided.

The 'Sendai' virus is a pleomorphic, but often roughly spherical, particle in which the nucleic acid (RNA) is surrounded by a lipo-protein envelope[39,40] (Fig. 1.2). The fusing ability of the virus resides in the viral envelope, not in the nucleic acid. The nucleic acid may be inactivated by large doses of ultraviolet light,[1] or destroyed by reaction with β-propiolactone,[41] without serious impairment of the fusing ability of the virus; and even fragments of viral envelope produced by ultrasonic disintegration of the virus retain to some extent the ability to induce cell fusion.[42] On the other hand, removal of lipid from the viral membrane by treatment with ether completely abolishes fusing ability.[43] In the related New-castle Disease virus, the ability to fuse has been shown to be sensitive to the action of phospholipase.[44] These observations suggest that fusion requires the structural integrity of the viral membrane, but some preliminary experiments have been done which suggest that it may be possible to extract from the 'Sendai' virus a protein that still has the ability to induce fusion.[45]

When a suspension of cells is treated with a high enough dose of virus to ensure a high multiplicity of particles per cell, the cells clump together. The size of the clumps is, within limits, proportional to the amount of virus added. What changes take place on the cell surface to determine this increased cellular adhesiveness are not clear. Clumping takes place within a few seconds at 4°C, so that it is perhaps unlikely that the changes in the cell surface are produced by enzymatic activity.

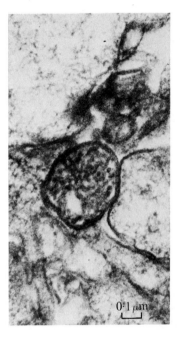

Fig. 1.2. A Sendai virus particle attached to the surface of a cell. A distinct outer membrane and internal coils of nucleoprotein are shown.

Electron micrographs of cells clumped together by the virus show the virus particles trapped between adjacent cell membranes or enmeshed in the microvilli that are found on the surfaces of many cells (Fig. 1.3). It is possible that the virus particle itself acts as the adhesive agent by adsorbing strongly to both of the apposed cell membranes with which it is in contact. A more precise analysis of the initial clumping reaction is at present frustrated by the uncertainties inherent in measuring

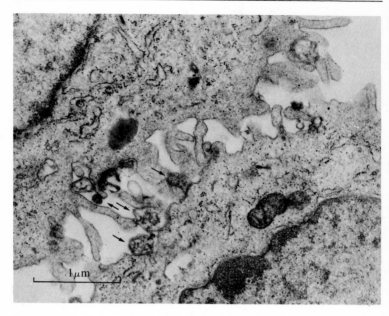

FIG. 1.3. Two adjacent cells showing virus particles (arrows) enmeshed in surface microvilli.

the efficiency of collision between virus particles and cells.[46] Although clumping takes place at 4°C, cell fusion does not begin until the temperature is raised. Okada, Murayama, and Yamada[47] have presented evidence in support of the view that cell fusion is an energy-requiring reaction inhibited by conditions that interfere with oxidative phosphorylation in the cell. Calcium ions also appear to be essential.[48] Since cells that multiply under virtually anaerobic conditions can be fused together by the virus, it seems unlikely that the fusion reaction can be intimately coupled to oxidative phosphorylation itself. Okada et al.[47,48] consider that the virus produces discontinuities in the cell membrane and that fusion between two cells occurs when the discontinuities in their surfaces are apposed. Electron microscopy has not so far resolved the precise structural basis of fusion. Hosaka and Koshi[49] consider that the virus particle itself forms the initial bridge between the two adjacent cells. This is a plausible scheme in the light of what is known about the mode of formation of the envelope of myxoviruses and the

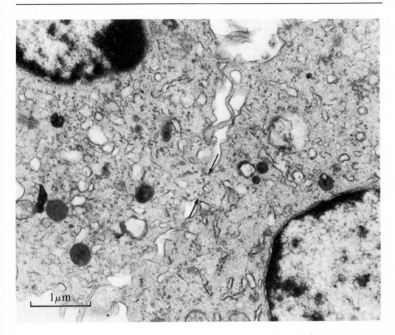

FIG. 1.4. Two adjacent cells showing the formation of a small cytoplasmic bridge.

mechanism by which these viruses enter the cell. It appears that myxoviruses receive their envelope as they pass out through the cell membrane,[50,51] and when the virus re-enters the cell, the viral envelope fuses back into the cell membrane and thus permits the entry of the viral RNA.[52,53] It is therefore reasonable to suppose that where the surfaces of two cells are closely apposed, the envelope of the virus particle might fuse with both cell surfaces and thus form an intercellular bridge. The earliest sign of cell fusion seen with the electron microscope is indeed the formation of minute cytoplasmic bridges between the adjacent surfaces of the apposed cells (Fig. 1.4). However, the evidence that the virus particles themselves form an essential part of all cytoplasmic bridges is not conclusive.[39] With time, the number of cytoplasmic bridges and the size of the individual bridges increase, so that eventually the cytoplasms of adjacent cells coalesce (Fig. 1.5). Fragments of cell membrane trapped in this coalescence are sometimes seen within the cytoplasm of the fused cell in the form of vesicles. The speed of fusion varies

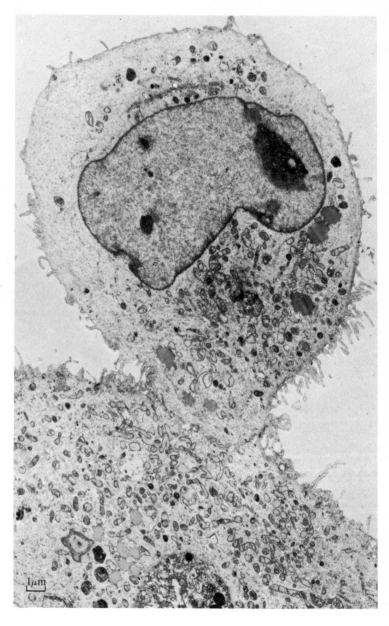

FIG. 1.5. The boundary between two cells in the process of fusing with each other.

with different cell types; with highly susceptible cells fusion may be complete in less than 5 min at 37°C.

The range of 'Sendai' virus

Although the ability of 'Sendai' virus to fuse like tumour cells together had been clearly established by the work of Okada and his colleagues,[27,28] it was not known prior to the work of Watkins and myself,[1] whether the virus could fuse unlike cells together. Nor was it known whether it could act on normal diploid cells or differentiated cells. Indeed, some preliminary studies by Okada and Tadokoro[54] were thought to show that the virus would not act on differentiated cells; and it was suggested that the ability of tumour cells to fuse under the action of the virus was in some way related to their neoplastic properties. Watkins and I were able to show that the virus would fuse together cells from widely different animal species; and a rapid screening of a range of normal differentiated diploid cells soon revealed that they, too, could be fused by the virus.[55] The range of species over which the virus will act has not yet been fully explored, but a variety of interspecific fusions has been achieved between human cells, cells from all the common laboratory animals, and cells from birds and frogs. It thus appears that interspecific cell fusion will be possible over a large part of the vertebrate subphylum. Fish cells have not yet been tried, but insect cells in culture do not appear to be susceptible to the virus.[56] Although the facility with which fusion can be effected varies greatly from one cell type to another, even the most highly differentiated cells appear to be capable of fusing under appropriate conditions. Fibroblasts, macrophages, lymphocytes, and nucleated erythrocytes have all been successfully fused both with each other and with other cells.[55,57,58] The facility with which fusion takes place appears to be related in some way to the structure of the cell surface. Cells with highly irregular surfaces, especially those whose surfaces are densely covered with microvilli, fuse most readily, whereas cells with relatively smooth and regular surfaces fuse with difficulty. This is perhaps the reason why tumour cells and cell lines grown *in vitro* tend to fuse more easily than most normal differentiated cells: a profusion of microvilli is characteristic of the surface of many tumour cells and of cell lines grown for long periods *in vitro*. It is possible that the difference between cells that fuse well and those that fuse poorly may be a reflection of the degree of contact that is possible between the adjacent cell surfaces. Where the cell surfaces are highly plastic, and especially where interdigitation of microvilli occurs, the effective areas of contact between two apposed cells must be very much greater than where two essentially spherical cells with smooth surfaces are in contact. More sophisticated differences in the nature of the cell surfaces are, of course, not excluded.

Multinucleate cells

The experiments that Watkins and I described contained one extension of the work of Okada which proved to be of overriding importance for the further development of cell hybridization: we showed that the multinucleate hybrid cells produced by the action of the virus were viable. In our original experiments we chose as the parent cells the HeLa cell, a line of human cells that has been maintained in tissue culture for many years, and the Ehrlich ascites cell, a tumour that grows in suspension in the peritoneal cavity of the mouse. We chose these cells for a variety of trivial technical reasons, but mainly because they had nuclei that were easily distinguishable on morphological grounds. This feature, we thought, would make it possible to detect at a glance whether the induced cell fusion involved both parent cells, and would permit us to examine the activity of each of the parent nuclei in the multinucleate hybrid cells, if these cells indeed survived.

When the mixed cell suspension that had been treated with the virus was introduced into culture chambers at 37°C, the fused cells adhered to and spread out on the floor of the chambers. Fixed and stained preparations revealed cells with varying numbers of nuclei containing in some cases nuclei recognizable as HeLa and nuclei recognizable as Ehrlich within the one cell (Fig. 1.6). There was thus every reason to believe that fusion had been effected between cells derived from two different species. But, in order to remove all possible doubt, a fusion experiment was carried out between HeLa cells in which the nuclei had previously been labelled with tritiated thymidine and Ehrlich cells that had not been labelled. Autoradiographs

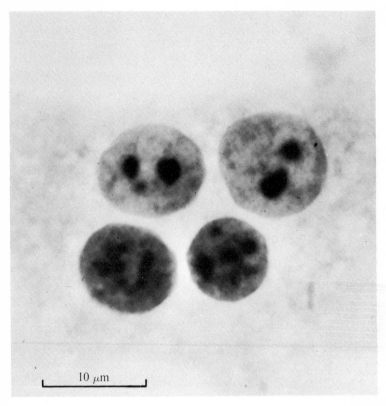

10 μm

FIG. 1.6. A tetranucleate cell in which the two upper nuclei are derived from HeLa cells and the two lower ones from Ehrlich ascites cells.

of the population of multinucleate cells formed in this way revealed many cells containing both labelled and unlabelled nuclei (Fig. 1.7). It was therefore clear than heterokaryons had been formed from parent cells of human and murine origin. A titration of the concentration of virus used against the average number of nuclei per cell in the resultant population of fused cells revealed that the extent of multinucleation could, within certain limits, be controlled by the dose of virus added to the cell suspension. The proportion of nuclei of each type in the multinucleated cells could also be controlled by varying the ratio of the two parent cells in the suspension. The use of the inactivated virus thus made available a precise method for

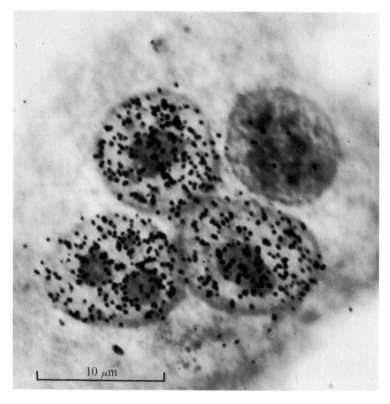

FIG. 1.7. Autoradiograph of a tetranucleate cell containing 3 HeLa nuclei and 1 Ehrlich nucleus. The HeLa cells had been grown in tritiated thymidine before the heterokaryons were produced. The HeLa nuclei are labelled and the Ehrlich nucleus is not.

producing, at will, interspecific heterokaryons of known composition.

Conventional radioactive techniques soon revealed that these heterokaryons synthesized protein and RNA, and autoradiographic procedures established that all the nuclei in the heterokaryon contributed to this RNA synthesis (Fig. 1.8). In these heterokaryons, therefore, the genes of both mouse and man were being transcribed. The heterokaryons also synthesized DNA, and it could again be shown by autoradiographic procedures that replication of DNA could occur in both sets of nuclei. But an examination of the timing of DNA synthesis in the individual nuclei in the heterokaryon revealed a situation which, on further analysis, proved to be of some interest.

When cells growing at random are fused together, the multi-nucleate cells are formed from parents at all stages of the cell

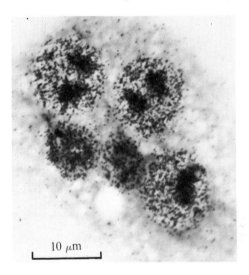

10 μm

FIG. 1.8. Autoradiograph of a heterokaryon containing 3 HeLa and 2 Ehrlich nuclei, exposed for 2 h to a radioactive RNA precursor. All the nuclei are synthesizing RNA.

cycle. This was established by an analysis of the proportions of labelled nuclei in multinucleate cells produced from a cell population that had been exposed to tritiated thymidine immediately before fusion.[59] Fusion does not appear to select either for or against any particular stage of the cell cycle (except possibly the period of mitosis itself). Immediately after fusion, therefore, multinucleate cells may contain nuclei that have not yet begun to synthesize DNA (G1 phase cells), nuclei at various stages in the period of DNA synthesis (S phase cells), and nuclei that have completed the replication of their DNA (G2 phase cells). Exposure of the multinucleated cells to tritiated thymidine directly after cell fusion therefore reveals an essentially random pattern of nuclear labelling. In multinucleate cells formed by the fusion of two or more HeLa cells, synchronization of DNA synthesis usually occurs rapidly.[59] Some degree of synchronization can be observed within 5 hours of fusion, and, within 24 hours, the nuclei in about 85 per cent of all such multinucleated cells are synchronized with respect to DNA synthesis. The main factor responsible for the imposition of synchrony appears to be the metabolic dominance of cells already in the S phase at the time of fusion. When such cells are fused with G1 phase cells, DNA synthesis is imposed on the quiescent partners. Once one of the nuclei in the multinucleate cell begins to synthesize DNA, the other nuclei rapidly follow in most cases. The persistent asynchronous DNA synthesis observed in about 15 per cent of the multinucleate cells has not yet been satisfactorily explained; it appears not to be due simply to the incorporation into the multinucleate cell of G2 phase nuclei.

When fusion is induced in a population of cells that has been artificially synchronized, the degree of co-ordination seen in the resultant multinucleate cells 24 hours after fusion is not any higher than that seen at the same time in multinucleate cells produced from a population growing at random.[59,60] This indicates that the use of virus to induce cell fusion does not impair the ability of the resulting multinucleate cells to co-ordinate nuclear events. The degree of synchrony and the speed with which it is imposed is, however, influenced by the number of nuclei in the multinucleate cell. Synchronization is rapidly achieved in the vast majority of binucleate cells, but, with increasing numbers of nuclei, asynchronous patterns of DNA synthesis become more common. In cells containing more than two nuclei, there is also a tendency for some measure of synchrony to be imposed transitorily, only to break down at a later stage.[59] These observations on fused HeLa cells are essentially similar to those made by Yamanaka and Okada[61] on fused KB cells (an established line of human cells). It is probable that rapid synchronization of the cell cycle will prove to be the general rule whenever like cells are fused together to form homokaryons.

In heterokaryons, where unlike cells are fused together, the position is more complicated. In some cases, synchrony is readily established,[62] but in others various degrees of asynchrony persist.[63] Persistent asynchrony of DNA synthesis was indeed observed in the very first heterokaryons that were produced, those made by fusing HeLa cells with Ehrlich ascites cells.[1] When these heterokaryons were exposed to tritiated

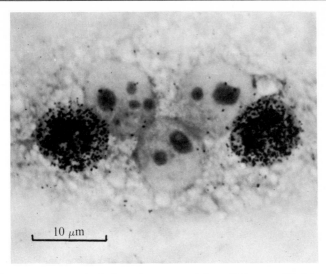

FIG. 1.9. Autoradiograph of a heterokaryon containing 3 HeLa and 2 Ehrlich nuclei, exposed for 2 h to a radioactive DNA precursor. The Ehrlich nuclei are labelled, but the HeLa nuclei are not.

thymidine, it was found that the Ehrlich nuclei were labelled with a much higher frequency than the HeLa nuclei (Fig. 1.9); and this disparity between the two sets of nuclei persisted for many days. A more precise analysis of this situation revealed a new phenomenon of considerable interest. It was found that the persistent asynchrony was due to competition between the Ehrlich nuclei and the HeLa nuclei in the same cell for some limiting factor or factors essential for the synthesis of DNA.[63] In this competition the Ehrlich nuclei were overwhelmingly successful. They behaved essentially as intracellular parasites, arrogating to themselves whatever was necessary for DNA synthesis at the expense of the HeLa nuclei, in which DNA synthesis was consequently largely inhibited. By varying the proportions of HeLa and Ehrlich nuclei in the heterokaryons this intracellular competition could be accurately titrated. Whether the ability to compete favourably for the prerequisites of DNA synthesis is a peculiarity of the Ehrlich nucleus, or whether this property is shared, in varying degrees, by the nuclei of other malignant cells is an important question which is now being examined. The HeLa–Ehrlich heterokaryon presents the most extreme degree of nuclear asynchrony which has so far been observed; but lesser degrees of asynchrony are commonly observed in other heterokaryons, especially where the physiological states of the nuclei of the two parent cells are very different.[62] Some interesting examples of this situation will be discussed at a later stage.

Nuclear fusion

At least in culture, animal cells do not appear to be able to multiply indefinitely in the multinucleate state. There are, of course, numerous examples in nature of syncytial forms of life in which growth and multiplication of the organism can take place without difficulty in the multinucleate state either by synchronous or asynchronous mitosis of the nuclei in the syncytium. But for these artificially produced multinucleate cells, and probably also for the multinucleate cells found in various conditions in the animal body, this mode of growth appears to be precluded. Although the multinucleate cell may, under favourable conditions, remain alive for several weeks, its continued reproduction is conditional upon the formation of daughter cells which contain a single nucleus. This process may be achieved in a variety of ways, but all of them are mediated by fusion of the individual nuclei in the multinucleate cell into larger units. This fusion takes place at mitosis.

One mechanism by which binucleate cells arising spontaneously in culture give rise to mononucleate daughters has already been described.[29,30] An essentially similar process may also occur in binucleate heterokaryons: the two nuclei enter mitosis together, a single spindle is formed, all the chromosomes become aligned along one metaphase plate, and cell division gives rise to mononucleate daughters which contain within a single nucleus the chromosomes of both parent cells. Figures 1.10–1.13 show frames from a cinemicrographic record of this form of mitosis in a HeLa–Ehrlich heterokaryon. By chromosomal analysis of subsequent metaphases, the mononucleate daughters can be shown to contain both the HeLa and the Ehrlich chromosomes. Other forms of mitosis are also seen, especially tripolar and tetrapolar mitosis, which give rise to variable numbers of daughter cells, some of which may be binucleate. In other cases, the nuclei in the heterokaryon may

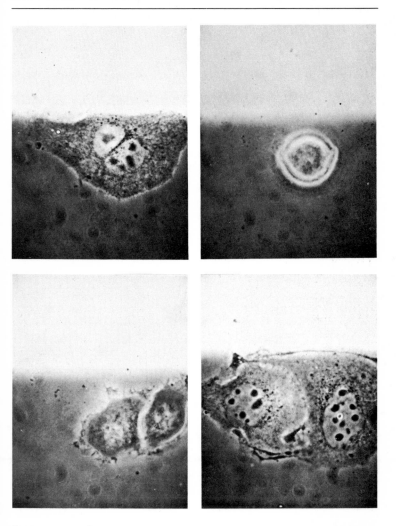

Figs. 1.10–1.13. Four frames from a cinematographic sequence showing a HeLa–Ehrlich heterokaryon undergoing mitosis and giving rise to two mononucleate daughter cells (synkaryons).

enter mitosis, but cell division may not occur. Post-mitotic re-constitution may then gather all the chromosomes of the cell into a single very large nucleus, or two nuclei may be formed each containing both the parental sets of chromosomes. There are thus several ways in which a binucleate heterokaryon can generate daughter cells in which a single nucleus contains genetic components from both parents. Such hybrid mono-nucleate daughters have been termed synkaryons.

Where cells contain more than two nuclei, irregular and abortive mitosis becomes increasingly common. Cells in which DNA synthesis is effectively synchronized usually undergo some form of mitosis, and their nuclei enter mitosis more or less together. [59] However, these multiple mitoses may be grossly irregular and they are often associated with failure to form any organized spindle at all. Post-mitotic reconstitution may then result in very bizarre nuclei and extensive micronucleation. The cell itself may not divide. In cells where asynchrony of DNA synthesis persists, the incidence of mitosis is greatly re-duced,[63] and when it does occur, the chromosomes of the lag-ging nucleus or nuclei may undergo extensive disruption by an ill-defined process which has been termed 'chromosome pul-verization'. The yield of viable mononucleate daughter cells is very low in heterokaryons which show persistent nuclear asynchrony.

Reproduction of synkaryons

While all synkaryons are able to synthesize RNA and DNA and in many cases undergo mitosis, not all of them are capable of continued multiplication. What kinds of restrictions there might be on the multiplication of these cells have not yet been fully explored. One obvious consideration is the chromosomal constitution of the synkaryon nucleus. Where the synkaryons are produced by an irregular mitotic event in the heterokaryon, they may fail to receive some of the chromosomes present in the parental nuclei, or they may inherit grossly unbalanced chromosomal sets. Such maldistribution of chromosomes could, of course, be lethal, although, as I shall mention later, sub-stantial chromosomal losses may occur in the hybrid cell with-out impairing its ability to multiply. Synkaryons containing more than two parental chromosomal sets may also be capable of multiplication. Virus-induced fusion between a line of hamster cells and a line of mouse cells has recently yielded a rapidly growing interspecific hybrid containing at least three parental chromosomal sets.[64] Extensive polyploidy obviously does not, in itself, constitute a barrier to vigorous multiplication. Nevertheless, the vast majority of hybrid cells that have been propagated contain only one chromosomal set from each parent, even when they are derived from populations of fused cells that contain a high average number of nuclei per cell. It is therefore clear that some factors do militate against the production or growth of hybrid cells of higher ploidy. One of these factors may well be the relative infrequency with which viable synkaryons are produced from heterokaryons that initially contain a large number of nuclei.

Over a wide range, species differences in the parent cells do not appear to affect the ability of synkaryons to multiply. Hybrids capable of indefinite multiplication have now been produced by virus-induced fusion between cells derived from mouse and hamster,[64] mouse and man,[65-68] and even, apparently, mouse and chick.[69] It is therefore unlikely that restrictions on multiplication are imposed by species differences alone. Nor does differentiation present an insuperable barrier. Interspecific hybrid cells have been propagated from heterokaryons produced by fusing mouse cells from an established line with leucocytes from human peripheral blood,[65,66,68] and from heterokaryons produced by fusing similar mouse cells with nucleated erythrocytes or leucocytes from the blood of erythroblastotic new-born babies.[70] Rapidly multiplying hybrids have also been produced by fusing together euploid human fibroblasts taken directly from two different individuals.[71,72] However, not all cell types are readily compatible. Although vigorous hybrids have resulted from the fusion of several different kinds of human and murine parent cells, synkaryons produced by fusing HeLa and Ehrlich cells multiply very slowly; and although small clones of cells can be derived from such synkaryons, a continuous cell line has not yet been produced.[73] What parental properties determine incompatibility in the hybrid cell is at present a totally unexplored field.

The growth rate of virus-induced synkaryons

Synkaryons show, relative to their parent cells, great variations in growth rate. Yamanaka and Okada,[74] using cinemicrography, found that when two KB cells were fused together with inactivated virus, the synkaryons formed at mitosis usually divided at least as rapidly as the parent cells. Engel, McGee, and

Harris[75] found that the hybrids produced by fusing mouse cells from two different established cell lines multiplied at about the same rate as the parent cells for about 3 weeks after cell fusion; but, after 3 months' cultivation, these hybrids easily outgrew the two parent cells in mixed culture. This presumably indicates that some of the variants generated by the hybrid cells were capable of more rapid growth than any of the variants generated by the parent cells. Yerganian and Nell[76] found that the hybrids produced by fusing near diploid cells derived from two different species of hamster at once outgrew the two parent cells. Hybrids produced by fusing together euploid fibroblasts taken directly from a patient with glucose-6-phosphate dehydrogenase deficiency and similar cells taken from another patient with inosinic acid pyrophosphorylase deficiency also outgrew the two parent cells.[72] Interspecific hybrids may also grow more rapidly than their parent cells even when the hybrids contain more than two sets of chromosomes.[64] A study of individual clones of hybrid cells, each derived from a primary fusion between an Ehrlich ascites cell and a mouse fibroblast from an established cell line, has shown that the same parent cells can generate hybrid clones of very different morphology and growth rate.[77] Some of the clones examined in this study contained cells that at once grew more rapidly than the parent cells, others contained cells that grew less rapidly. The parent cells in this case were aneuploid, so that the clonal variation might have had its basis in differences in chromosomal content. On the other hand, as mentioned previously, hybrids produced by fusing Ehrlich ascites cells with HeLa cells invariably grew more slowly than the parent cells.[73] It thus appears that no general principles have yet emerged which would permit us to predict the growth rate of any particular hybrid; but it is clear that 'hybrid vigour', as defined by increased growth rate, is a property of some virus-induced hybrids, both intraspecific and interspecific.

'Spontaneous' versus virus-induced hybrids

Before the publication of the Harris and Watkins experiment, cells produced by 'spontaneous' hybridization had only been sought in mixed cultures of different kinds of mouse cells; but, shortly afterwards, Ephrussi and Weiss[78] reported the spontaneous occurrence of a rat–mouse hybrid cell in mixed cultures of rat and mouse cells. Using, for the most part, the selective procedures devised by Littlefield,[37] Ephrussi and his colleagues[79–81] and other authors[82,83] have since isolated a number of interspecific hybrid cells arising spontaneously in mixed cultures of the parents. Neither the intraspecific nor the interspecific hybrid cells produced in this way differ in any important particular from those produced deliberately by the use of inactivated virus. The studies of Littlefield[37,84] and those of Engel *et al.*[75] permit a comparison to be made between spontaneous and virus-induced hybrids from the same parent cells. The parent cells in both studies were Littlefield's A9 and B82 strains of mouse fibroblast, one of which lacks inosinic acid pyrophosphorylase, the other thymidine kinase.[84] Hybrid cells contain both enzymes and can thus be selected for in an appropriate medium.[37] In the experiments of Engel *et al.*,[75] the two cell types were fused together by means of inactivated virus; in Littlefield's experiments spontaneous hybrid cells were selected from mixed cultures.[84] The growth potential, the initial chromosomal constitution, and the subsequent behaviour of the hybrid cells were essentially the same in the two sets of experiments. The yield of hybrid clones was, however, greatly increased by the use of the inactivated virus. A similar observation was made by Siniscalco[72] in his study of the hybrid cells produced by fusing together human fibroblasts with different metabolic defects. Again, the yield of hybrid cells was greatly increased by the use of inactivated virus, but in their growth rate, chromosomal constitution, and metabolic activity the virus-induced hybrids and the spontaneous hybrids were indistinguishable. It appears that if two cell types will yield viable hybrids when cultivated together, they will do so much more readily when fused together by virus. In the design of cell hybridization experiments there would therefore seem to be no reason for relying on the unpredictable and rare fusion events that occur spontaneously in mixed cultures. The use of virus imposes cell fusion under controlled conditions and greatly increases the yield of viable hybrid clones. But the outstanding advantage of the virus technique is that it will induce hybridization between cells that cannot fuse together spontaneously. It is this that converts cell hybridization from an exercise in the exploitation of chance events to a method of general applicability.

The chromosomes of hybrid cells and the possibilities for genetic analysis

The great majority of the hybrid cells that have been propagated appear initially to have contained one set of chromosomes from each parent cell (Figs. 1.14–1.16).[31,36,72,75,76,81,84,85]

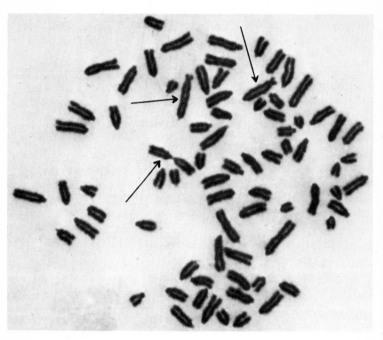

FIG. 1.14. The chromosomes of the Ehrlich cell. Marker chromosomes are indicated by arrows.

Where the parent cells are closely related, and, apparently, in all intraspecific combinations that have so far been studied, the karyotype of the hybrid cell is remarkably stable. Continued cultivation does result in a progressive loss of chromosomes, but at a very slow rate.[36] Littlefield[84] found that the hybrids that arose spontaneously in mixed cultures of the A9 and B82 cell lines lost an average of only 9 chromosomes out of a total of about 100 after 1 year's cultivation. Engel *et al.*[75] found that the average number of chromosomes in hybrids produced from the same two cell lines by virus-induced fusion remained essentially constant for the first 3 months of cultivation. Yoshida and Ephrussi[85] found a similarly slow and unpredictable loss of chromosomes in seven different kinds of mouse cell hybrid.

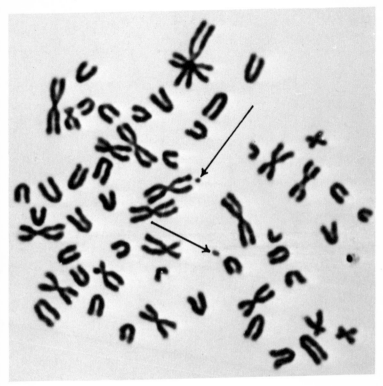

Fig. 1.15. The chromosomes of the A_9 cell. Marker 'dot' chromosomes are indicated by arrows.

However, although loss of chromosomes is not a striking feature of intraspecific hybrids, minor structural changes take place continuously and can be detected quite early. In the A9/B82 hybrids studied by Engel *et al.*,[75] chromosomes with morphological features not present in either of the parent cells were already obvious within 1 month after cell fusion. New ring chromosomes, dicentric chromosomes, and chromosomal fragments appeared in increasing numbers during the first 3 months of cultivation, and chromosomes engaged in strand inter-

C

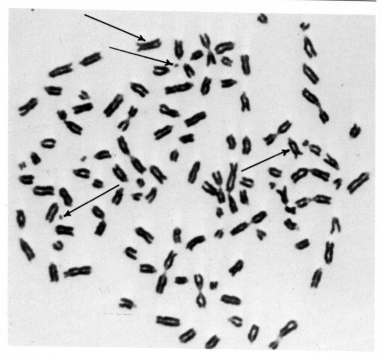

FIG. 1.16. The chromosomes of an A_9-Ehrlich hybrid cell. Marker chromosomes from both parents are indicated by arrows.

changes were also seen. The karyotype of these hybrids was thus far from static even though no net loss of chromosomes was detected during these first 3 months.

Littlefield[84] noted in his original studies on the A9/B82 hybrids that occasional cells with very low numbers of chromosomes appeared in the hybrid cultures. Engel *et al.*[75] observed the frequent occurrence of nuclear budding and nuclear fragmentation in these same hybrids; and Miller and Harris[86] found that populations of hybrids produced by the fusion of Ehrlich ascites cells with A9 cells contained occasional metaphase figures with as few as 20 chromosomes when the average number of chromosomes in these hybrids was about 130. These observations indicate that more radical segregation of chromosomes can occur in intraspecific hybrid cells. When the parent cells, like the A9 and B82 cells, have specific biochemical defects, some of these segregants can be studied because cultural conditions can be devised to select for them. Engel, McGee, and Harris[87] have shown that A_9/B_{82} hybrids grown continuously in selective medium for many months give rise to increasing numbers of segregants with a karyotype very similar to that of the parent cells; but these segregants show complementation of the parental enzymatic defects. The redundant genetic material in the hybrid cells appears to be progressively eliminated so that something very like the parental chromosomal complement is eventually restored. Where the parent cells do not have specific biochemical defects, the absence of any selection system for segregants has precluded a more extensive study of them. One important example of segregation that can be selected for *in vivo* will be discussed later in connection with experiments in which malignancy has been analysed by cell fusion.

The slow and unpredictable loss of chromosomes in intraspecific hybrid cells and the general inaccessibility of their more radical segregants hardly encouraged the hope that these hybrids would prove to be particularly useful in furthering the formal genetic analysis of somatic cells. But when interspecific hybrids began to be studied, the outlook was transformed. A strain of hybrid cells selected by Weiss and Green[82] from a mixed culture of diploid human embryonic lung fibroblasts and aneuploid mouse fibroblasts with a thymidine kinase deficiency eliminated the human chromosomes preferentially: in some cells almost all the human chromosomes were eliminated within twenty cell generations. This situation opened the possibility of correlating retention or loss of particular human chromosomes in the hybrid with retention or loss of specific phenotypic characters. On the basis of such a correlation, Weiss and Green[82] assigned the gene for thymidine kinase to one of the morphological subgroups of the human chromosomal complement. An approach to the mapping of human chromosomes thus appeared to be at hand. However, a more protracted analysis of a hybrid strain produced by fusing together cells from a patient with inosinic acid pyrophosphorylase deficiency and mouse cells with thymidine kinase deficiency revealed that thymidine kinase activity persisted in the hybrid cells even when all the human chromosomes had apparently been eliminated.[67] This finding illustrates two hazards inherent in genetic mapping by this means, one trivial, the other of more general importance. The trivial hazard is to assume that an enzyme deficiency induced in tissue culture cells by extreme selective procedures necessarily involves a mutation in the structural gene for that enzyme. We do not know that the absence or low level of thymidine kinase or inosinic acid pyrophosphorylase in these defective mouse cell lines are due to mutations in the genes coding for these enzymes; and unless we know this we cannot attempt to map the corresponding genetic loci. The second hazard is made clear by the study of Engel *et al.*[75] Minor chromosomal rearrangements take place constantly, even in intraspecific hybrids; so the possibility must always be borne in mind that the gene one is studying may have been translocated. Nevertheless, a solid confirmation of the X-linkage of the inosinic acid pyrophosphorylase gene in man has been obtained by analysis of man-mouse hybrid cells produced by fusing human leucocytes with mouse cells deficient in this enzyme; and evidence has also been obtained from the same cells that the human lactic dehydrogenase A and B genes are not linked.[65,66]

Rapid elimination of chromosomes is not limited to man-mouse hybrid cells. In the mouse–hamster hybrids which initially contained at least three parental chromosomal sets, some 30 chromosomes out of a total of about 130 were apparently lost within a month.[64] The factors responsible for rapid loss of chromosomes and the mechanisms by which this loss occurs are at present completely obscure; but there is evidence that the major losses probably occur during the early cell divisions.[66] It is clear that interspecific hybrid cells do, in principle, provide a new technique for genetic mapping; but the technique will need to be used with care.

Another interesting approach is the use of cell fusion for complementation analysis. Kao, Johnson, and Puck[88] have used the 'Sendai' virus technique to produce hybrids between mutant hamster cells with different nutritional requirements. Analysis of complementation in these hybrids permitted further elucidation of the mutations involved.

Conclusion

The most important conclusion to be drawn from these experiments is the fact that cells from different species of vertebrate are compatible with each other when they are amalgamated into a single unit. It thus appears that in the cells of vertebrates there are, in general, no *intracellular* mechanisms for the recognition of incompatibility similar to those responsible for the recognition and destruction of tissue or organ grafts exchanged between different individuals. Not only do the cytoplasms of these different cells fuse amicably together, but their nuclei also; and after nuclear fusion has taken place the composite cell carries out its functions in a perfectly integrated way, and may, in some cases, undergo vigorous and indefinite multiplication. These fused cells, both in their initial multinucleate, and in their subsequent mononucleate, state have lent themselves to experiments which, a few years ago, would not have been thought possible. Some of the more interesting of these experiments I shall discuss in the two subsequent chapters.

REFERENCES

1. HARRIS, H. and WATKINS, J. F. (1965). Hybrid cells derived from mouse and man: artificial heterokaryons of mammalian cells from different species. *Nature, Lond.* **205**, 640.
2. MÜLLER, J. (1838). *Ueber den feineren Bau und die Formen der krankhaften Geschwülste.* Berlin. Quoted by Faber.[3]
3. FABER, K. (1893). Multinucleate cells as phagocytes. *J. Path. Bact.* **1**, 349.
4. ROBIN, C. (1849). Sur l'existence de deux espèces nouvelles d'éléments anatomiques qui se trouvent dans le canal médullaire des os. *C. r. Séanc. Soc. Biol.*, p. 149.
5. ROKITANSKY, C. (1855). *Lehrbuch der pathologischen Anatomie,* 3rd edn, Vol. 1, p. 295. Braumüller, Vienna.
6. VIRCHOW, R. (1858). Reizung und Reizbarkeit. *Virchows Arch. path. Anat. Physiol.* **14**, 1.
7. LANGHANS, T. (1868). Ueber Riesenzellen mit wandständigen Kernen in Tuberkeln und die fibröse Form des Tuberkels. *Virchows Arch. path. Anat. Physiol.* **42**, 382.
8. LANGHANS, T. (1870). Beobachtungen über Resorption der Extravasate und Pigmentbildung in denselben. *Virchows Arch. path. Anat. Physiol.* **49**, 66.
9. BARY, A. DE (1859). Die Mycetozoen. Ein Beitrag zur Kenntnis der niedersten Thiere. *Z. wiss. Zool.* **10**, 88.
10. LANGE, O. (1875). Ueber die Entstehung der blutkörperhaltigen Zellen und die Metamorphosen des Blutes im Lymphsack des Frosches. *Virchows Arch. path. Anat. Physiol.* **65**, 27.

11. CIENKOWSKI, L. (1876). Ueber, einige Rhizopoden und verwandte Organismen. *Arch. mikrosk. Anat. EntwMech.* **12**, 15.

12. BUCK, E. (1878). Einige Rhizopodenstudien. *Z. wiss. Zool.* **30**, 1.

13. GEDDES, P. (1880). On the coalescence of amoeboid cells into plasmodia. *Proc. R. Soc.* **30**, 252.

14. METCHNIKOFF, E. (1884). Untersuchungen über die intracelluläre Verdauung bei wirbellosen Thieren. *Arb. zool. Inst. Univ. Wien.* **5**, 141.

15. LUGINBÜHL, D. (1873). Der Micrococcus der Variola. *Arbeiten aus dem Berner Pathologischen Institut,* 1871-2, p. 159. Würzburg.

16. WEIGERT, C. (1874). *Anatomische Beiträge zur Lehre von den Pocken,* p. 40. Breslau. Quoted by Krauss (1884).[17]

17. KRAUSS, E. (1884). Beiträge zur Riesenzellenbildung in epithelialen Geweben. *Virchows Arch. path. Anat. Physiol.* **95**, 249.

18. UNNA, P. G. (1896). *The histopathology of the diseases of the skin,* p. 637. Clay, Edinburgh.

19. WARTHIN, A. S. (1931). Occurrence of numerous large giant cells in the tonsils and pharyngeal mucosa in the prodromal stage of measles. *Archs Path.* **11**, 864.

20. HARRISON, R. G. (1907). Observations on the living developing nerve fibre. *Proc. Soc. exp. Biol. Med.* **4**, 140.

21. LAMBERT, R. A. (1912). The production of foreign body giant cells *in vitro. J. exp. Med.* **15**, 510.

22. LEWIS, W. H. (1927). The formation of giant cells in tissue cultures and their similarity to those in tuberculous lesions. *Am. Rev. Tuberc. pulm. Dis.* **15**, 616.

23. ENDERS, J. F. and PEEBLES, T. C. (1954). Propagation in tissue cultures of cytopathogenic agents from patients with measles. *Proc. Soc. exp. Biol. Med.* **86**, 277.

24. HENLE, G., DEINHARDT, F., and GIRARDI, A. (1954). Cytolytic effects of mumps virus in tissue cultures of epithelial cells. *Proc. Soc. exp. Biol. Med.* **87**, 386.

25. CHANOCK, R. M. (1956). Association of a new type of cytopathogenic myxovirus with infantile croup. *J. exp. Med.* **104**, 555.

26. MARSTON, R. Q. (1958). Cytopathogenic effects of haemadsorption virus Type 1. *Proc. Soc. exp. Biol. Med.* **98**, 853.

27. OKADA, Y. (1958). The fusion of Ehrlich's tumor cells caused by H.V.J. virus *in vitro. Biken's J.* **1**, 103.

28. OKADA, Y. (1962). Analysis of giant polynuclear cell formation caused by H.V.J. virus from Ehrlich's ascites tumor cells. *Expl. Cell Res.* **26**, 98.

29. MACKLIN, C. C. (1916). Binucleate cells in tissue culture. *Contr. Embryol.* **13**, 69. Carnegie Institute of Washington Publication 224.

30. FELL, H. B. and HUGHES, A. F. (1949). Mitosis in the mouse: a study of living and fixed cells in tissue cultures. *Q. Jl microsc. Sci.* **90**, 355.

31. BARSKI, G., SORIEUL, S., and CORNEFERT, F. (1960). Production dans des cultures *in vitro* de deux souches cellulaires en association, de cellules de caractère 'hybride'. *C. r. hebd. Séanc. Acad. Sci., Paris* **251**, 1825.

32. BARSKI, G. and BELEHRADEK, J. (1963). Transfert nucléaire intercellulaire en cultures mixtes *in vitro. Expl Cell Res.* **29**, 102.

33. BARSKI, G. (1968). Report on the workshop on virus induction by cell association. *Int. J. Cancer* **3**, 320.

34. SORIEUL, S. and EPHRUSSI, B. (1961). Karyological demonstration of hybridization of mammalian cells *in vitro. Nature, Lond.* **190**, 653.

35. EPHRUSSI, B. and SORIEUL, S. (1962). Nouvelles observations sur l'hybridation *in vitro* de cellules de souris. *C. r. hebd. Séanc. Acad. Sci., Paris* **254**, 181.

36. EPHRUSSI, B., SCALETTA, L. J., STENCHEVER, M. A., and YOSHIDA, M. C. (1964). Hybridization of somatic cells *in vitro*. In *Cytogenetics of cells in culture* (ed. R. J. C. Harris), p. 13. Academic Press, New York and London.

37. LITTLEFIELD, J. W. (1964). Selection of hybrids from matings of fibroblasts *in vitro* and their presumed recombinants. *Science, N.Y.* **145**, 709.

38. OKADA, Y. and TADOKORO, J. (1962). Analysis of giant polynuclear cell formation caused by H.V.J. virus from Ehrlich's ascites tumor cells. *Expl Cell Res.* **26**, 108.

39. SCHNEEBERGER, E. E. and HARRIS, H. (1966). An ultrastructural study of interspecific cell fusion induced by inactivated Sendai virus. *J. Cell Sci.* **1**, 401.

40. HOSAKA, Y., KITANO, H., and IKEGUCHI, S. (1966). Studies on the pleomorphism of HVJ virions. *Virology* **29**, 205.

41. NEFF, J. M. and ENDERS, J. F. (1968). Poliovirus replication and cytopathogenicity in monolayer hamster cell cultures fused with beta propiolactone-inactivated 'Sendai' virus. *Proc. Soc. exp. Biol. Med.* **127**, 260.

42. OKADA, Y. and MURAYAMA, F. (1968). Fusion of cells by HVJ: requirement of concentration of virus particles at the site of contact of two cells for fusion. *Expl Cell Res.* **52**, 34.

43. WATKINS, J. F. (1965). Unpublished result.

44. KOHN, A. and KLIBANSKY, C. (1967). Studies on the inactivation of cell-fusing property of Newcastle Disease virus by phospholipase A. *Virology* **31**, 385.

45. JOHNSON, R. T. (1968). Personal communication.

46. OGSTON, A. G. (1963). On uncertainties inherent in the determination of the efficiency of collision between virus particles and cells. *Biochim. biophys. Acta* **66**, 279.

47. OKADA, Y., MURAYAMA, F., and YAMADA, K. (1966). Requirement of energy for the cell fusion reaction of Ehrlich ascites tumor cells by HVJ. *Virology* **28**, 115.

48. OKADA, Y. and MURAYAMA, F. (1966). Requirement of calcium ions for the cell fusion reaction of animal cells by HVJ. *Expl Cell Res.* **44**, 527.

49. HOSAKA, Y. and KOSHI, Y. (1968). Electron microscopic study of cell fusion by HVJ virions. *Virology* **34**, 419.

50. WYCKOFF, R. W. G. (1951). Electron microscopy of chick embryo membrane infected with PR-8 Influenza. *Nature, Lond.* **168**, 651.

51. WYCKOFF, R. W. G. (1953). Formation of the particles of influenza virus. *J. Immun.* **70**, 187.

52. HOYLE, L. (1962). The entry of myxoviruses into the cell. *Cold Spring Harb. Symp. quant. Biol.* **27**, 113.

53. MORGAN, C. and ROSE, H. M. (1968). Structure and development of viruses as observed in the electron microscope. VIII Entry of influenza virus. *J. Virol.* **2**, 925.

54. OKADA, Y. and TADOKORO, J. (1963). The distribution of cell fusion capacity among several cell strains or cells caused by HVJ. *Expl Cell Res.* **32**, 417.

55. HARRIS, H. (1965). Behaviour of differentiated nuclei in heterokaryons of animal cells from different species. *Nature, Lond.* **206**, 583.

56. WATKINS, J. F. Personal communication.

57. HARRIS, H., WATKINS, J. F., FORD, C. E., and SCHOEFL, G. I. (1966). Artificial heterokaryons of animal cells from different species. *J. Cell Sci.* **1**, 1.

58. HARRIS, H., SIDEBOTTOM, E., GRACE, D. M., and BRAMWELL, M. E. (1969). The expression of genetic information: a study with hybrid animal cells. *J. Cell Sci.* **4**, 499.

59. JOHNSON, R. T. and HARRIS, H. (1969). DNA synthesis and mitosis in fused cells. 1. HeLa homokaryons. *J. Cell Sci.* In press.

60. WESTERVELD, A. (1969). Personal communication.

61. YAMANAKA, T. and OKADA, Y. (1966). Cultivation of fused cells resulting from treatment of cells with HVJ. I Synchronization of the stages of DNA synthesis of nuclei involved in fused multinucleated cells. *Biken's J.* **9**, 159.

62. JOHNSON, R. T. and HARRIS, H. (1969). DNA synthesis and mitosis in fused cells. 2. HeLa-chick erythrocyte heterokaryons. *J. Cell Sci.* In press.

63. JOHNSON, R. T. and HARRIS, H. (1969). DNA synthesis and mitosis in fused cells. 3. HeLa–Ehrlich heterokaryons. *J. Cell Sci.* In press.

64. WATKINS, J. F. (1969). Personal communication.

65. MIGGIANO, V., NABHOLZ, M., and BODMER, W. (1969). Hybrids between human leucocytes and a mouse cell line: production and characterization. *Proceedings of the Wistar Institute Symposium on Heterospecific Genome Interaction.* In press.

66. NABHOLZ, M., MIGGIANO, V., and BODMER, W. (1969). Genetic analysis using human-mouse somatic cell hybrids. *Nature, Lond.* **223**, 358.

67. MIGEON, B. R. and MILLER, C. S. (1968). Human-mouse somatic cell hybrids with single human chromosome (Group E): link with thymidine kinase activity. *Science N.Y.* **162**, 1005.

68. MILLER, O. J. (1969). Personal communication.

69. NABHOLZ, M. (1969). Personal communication.

70. MILLER, O. J. (1969). Personal communication.

71. SINISCALCO, M., KLINGER, H. P., EAGLE, H., KOPROWSKI, H., FUJIMOTO, W. F., and SEEGMILLER, J. E. (1969). Evidence for intergenic complementation in hybrid cells derived from two human diploid strains each carrying an X-linked mutant. *Proc. natn. Acad. Sci. U.S.A.* **62**, 793.

72. SINISCALCO, M. (1969). Hybridization of human diploid strains carrying X-linked mutants and its potentials for studies of somatic cell genetics. *Proceedings of the Wistar Institute Symposium on Heterospecific Genome Interaction.* In press.

73. WATKINS, J. F. and GRACE, D. M. (1967). Studies on the surface antigens

of interspecific mammalian cell heterokaryons. *J. Cell Sci.* **2**, 193.

74. YAMANAKA, T. and OKADA, Y. (1968). Cultivation of fused cells resulting from treatment of cells with HVJ. II Division of binucleated cells resulting from fusion of two KB cells by HVJ. *Expl Cell Res.* **49**, 461.

75. ENGEL, E., McGEE, B. J., and HARRIS, H. (1969). Cytogenetic and nuclear studies on A_9 and B_{82} cells fused together by Sendai virus. The early phase. *J. Cell Sci.* **5**, 93.

76. YERGANIAN, G. and NELL, M. B. (1966). Hybridization of dwarf hamster cells by UV-inactivated Sendai virus. *Proc. natn. Acad. Sci. U.S.A.* **55**, 1066.

77. HARRIS, H. (1969). In preparation.

78. EPHRUSSI, B. and WEISS, M. C. (1965). Interspecific hybridization of somatic cells. *Proc. natn. Acad. Sci. U.S.A.* **53**, 1040.

79. WEISS, M. C. and EPHRUSSI, B. (1966). Studies of interspecific (rat × mouse) somatic hybrids. I Isolation, growth and evolution of the karyotype. *Genetics, N.Y.* **54**, 1095.

80. SCALETTA, L. J., RUSHFORTH, N. B., and EPHRUSSI, B. (1967). Isolation and properties of hybrids between somatic mouse and Chinese hamster cells. *Genetics, N.Y.* **57**, 107.

81. EPHRUSSI, B. and WEISS, M. C. (1967). Regulation of the cell cycle in mammalian cells: inferences and speculations based on observations of interspecific somatic hybrids. In *Control mechanisms in developmental processes* (ed. M. Locke), Society for Developmental Biology Symposium No. 26. *Devl. Biol.* Suppl. I. p. 136.

82. WEISS, M. C. and GREEN, H. (1967). Human–mouse hybrid cell lines containing partial complements of human chromosomes and functioning human genes. *Proc. natn. Acad. Sci. U.S.A.* **58**, 1104.

83. CARVER, D. H., SETO, D. S. Y., and MIGEON, B. R. (1968). Interferon production and action in mouse, hamster and somatic hybrid mouse–hamster cells. *Science, N.Y.* **160**, 558.

84. LITTLEFIELD, J. W. (1966). The use of drug-resistant markers to study the hybridization of mouse fibroblasts. *Expl Cell Res.* **41**, 190.

85. YOSHIDA, M. C. and EPHRUSSI, B. (1967). Isolation and karyological characteristics of seven hybrids between somatic mouse cells *in vitro*. *J. cell. Physiol.* **69**, 33.

86. MILLER, O. J. and HARRIS, H. Unpublished result.

87. ENGEL, E., McGEE, B. J., and HARRIS, H. (1969). Recombination and segregation in somatic cell hybrids. *Nature, Lond.* **223**, 152.

88. KAO, F.-T., JOHNSON, R. T., and PUCK, T. T. (1969). Complementation analysis on virus-fused Chinese hamster cells with nutritional markers. *Science N.Y.* **164**, 312.

2. Gene activity and differentiation

Heterokaryons made with differentiated cells

DURING the course of differentiation certain vertebrate cells lose, in varying degrees, the ability to synthesize DNA or RNA or both. The demonstration that the membranes of these highly specialized cells were also susceptible to the action of 'Sendai' virus[1,2] opened up the possibility of examining certain aspects of differentiation in a way that had not previously been feasible. Three cells were initially chosen for special study: the rabbit macrophage, the rat lymphocyte, and the hen erythrocyte. Macrophages are motile, phagocytic cells, whose main function is the removal of débris from the tissues of the body. These cells can be obtained in large numbers from the peritoneal cavity of the rabbit after certain experimental procedures.[3] They commonly have an oval or kidney-shaped nucleus. The macrophages that one obtains from the peritoneal cavity of the rabbit all synthesize RNA, but they do not, either in the peritoneal cavity or *in vitro*, synthesize DNA or undergo mitosis.[4] The lymphocyte is a small cell with a dense compact nucleus and very little cytoplasm. In the rat, almost pure populations of lymphocytes can be obtained by cannulation of the thoracic duct. These small lymphocytes synthesize variable amounts of RNA: when exposed for an hour to high concentrations of a radioactive RNA precursor, some of the cells synthesize so little RNA that the amount of radioactivity incorporated is barely detectable, while other cells incorporate easily measurable amounts of radioactivity under the same conditions. Small lymphocytes do not normally synthesize DNA or undergo mitosis; but they can be induced to resume the synthesis of DNA and to undergo mitosis when they are exposed to certain antigenic stimuli.[5,6] The small lymphocyte then becomes transformed into a cell that plays a crucial role in the immune responses of the body. Whereas mammalian erythrocytes normally eliminate their nuclei during the process of maturation, the red blood cells of birds, amphibians, reptiles, and certain other orders of vertebrate retain their nuclei throughout the life-cycle of the cell. In the hen, these nucleated erythrocytes, when mature, do not synthesize measurable amounts of RNA, nor do they synthesize DNA or undergo mitosis. They are thus 'end-cells': after a variable period of circulation in the blood they are removed and destroyed. It was of interest in the first instance to determine, especially in the case of the nucleated erythrocyte, to what extent these changes produced by differentiation were reversible. Could the inert or partially inactive cell be induced to resume the synthesis of RNA or DNA or both, if it were incorporated in a heterokaryon together with a cell which synthesized both RNA and DNA in the normal way?

The regulation of nucleic acid synthesis

This problem was first investigated by fusing each of these differentiated cells with HeLa cells.[1,2] Figures 2.1–2.3 show examples of the heterokaryons produced. By varying the ratio of the two cell types used and the concentration of inactivated virus, the average number of nuclei per heterokaryon and the proportion of each kind of nucleus could, over a certain range, be controlled. It was thus possible to produce heterokaryons in which a single HeLa cell had been fused with several highly differentiated cells (Fig. 2.1). Autoradiographic experiments with appropriate radioactive precursors rapidly revealed that, in these heterokaryons, irrespective of the number of differentiated nuclei present, the macrophage and the lymphocyte nuclei resumed the synthesis of DNA (Fig. 2.4), and the erythrocyte nuclei resumed the synthesis of both RNA and DNA (Fig. 2.5). The restrictions on nucleic acid synthesis which are normally operative in these highly differentiated cells are thus removed when these cells are fused with a HeLa cell. And since, among nucleated cells, erythrocytes represent perhaps the most extreme form of differentiation seen in vertebrates, one is probably justified in supposing that, so long as the cell

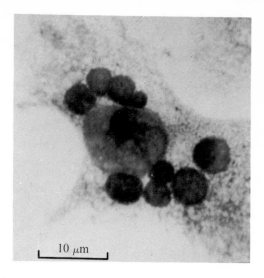

FIG. 2.1. A heterokaryon containing one HeLa nucleus and a number of rabbit macrophage nuclei.

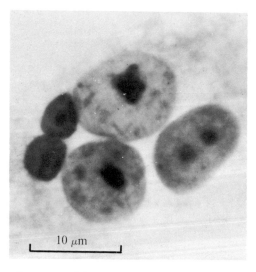

FIG. 2.2. A heterokaryon containing three HeLa nuclei and two rat lymphocyte nuclei.

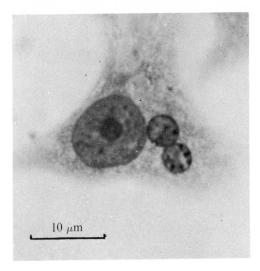

FIG. 2.3. A heterokaryon containing one HeLa nucleus and two chick erythrocyte nuclei.

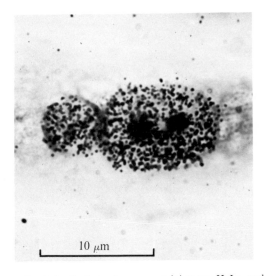

FIG. 2.4. Autoradiograph of a heterokaryon containing one HeLa nucleus and one rabbit macrophage nucleus, exposed for 2 h to a radioactive DNA precursor. Both nuclei are synthesizing DNA.

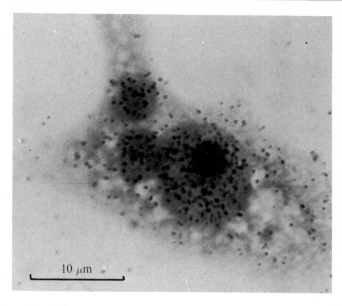

Fig. 2.5. Autoradiograph of a heterokaryon containing one HeLa nucleus and two chick erythrocyte nuclei exposed for 4 h to a radioactive RNA precursor. The HeLa nucleus and the erythrocyte nuclei are synthesizing RNA.

retains its nucleus, all restrictions on nucleic acid synthesis imposed by the process of differentiation are reversible.

The reactivation of these nuclei can be further analysed by fusing the differentiated cells with each other. It could be argued, for example, that nucleic acid synthesis is resumed in these nuclei in the heterokaryon, not because one of the parent cells from which the heterokaryon was formed was itself active in this respect, but simply because the dormant nuclei now find themselves in a foreign environment. It could be supposed that the reactivation of these nuclei is a non-specific response to 'foreignness'. However, analysis of heterokaryons in which rabbit macrophages are fused with rat lymphocytes (Fig. 2.6) or hen erythrocytes (Fig. 2.7) shows that this is not the case.[2,7] In these heterokaryons the nuclei are in foreign, or partially foreign, cytoplasm, in that the heterokaryons are produced from cells of different species, but these nuclei do not react in the same way as they do when the differentiated cells are fused with HeLa cells. In macrophage–lymphocyte heterokaryons,

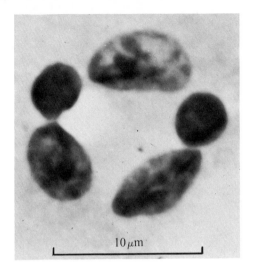

Fig. 2.6. A heterokaryon containing three rabbit macrophage nuclei and two rat lymphocyte nuclei, which are smaller and stain more deeply. Note the peripheral distribution of the nuclei in the cell.

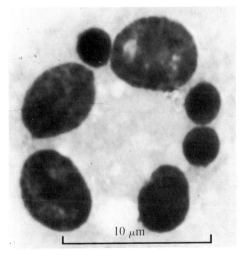

Fig. 2.7. A heterokaryon containing four rabbit macrophage nuclei and three hen erythrocyte nuclei, which are much smaller. Note the peripheral distribution of the nuclei in the cell.

where both of the parent cells normally synthesize RNA, but not DNA, the nuclei in the heterokaryon also synthesize RNA, but not DNA; and in macrophage–erythrocyte heterokaryons, the erythrocyte nuclei are again induced to synthesize RNA, but neither the erythrocyte nor the macrophage nuclei in the heterokaryon synthesize DNA. On the other hand, if chick erythrocyte nuclei are introduced into the cytoplasm of fibroblasts grown from the same chick embryo, the erythrocyte nuclei in the heterokaryons resume the synthesis of both RNA and DNA, even though they are not now in genetically foreign cytoplasm.[8,9] It is therefore clear that it is not foreignness that determines the reactivation of these dormant nuclei in heterokaryons, but the metabolic state of the cells from which the heterokaryons are formed. It is also clear that the ability to reactivate these nuclei is not a peculiarity of the HeLa cell or of other aneuploid cells long established in culture; normal diploid cells can also achieve this end.

The results of these experiments with differentiated cells are summarized in Table 2.1, from which it will be seen that certain general principles emerge: (1) if either of the parent cells normally synthesizes RNA, then RNA synthesis will take place in both types of nuclei in the heterokaryon; (2) if either of the

TABLE 2.1
Synthesis of RNA and DNA in heterokaryons

	RNA	DNA
Cell type		
HeLa	+	+
Rabbit macrophage	+	○
Rat lymphocyte	+	○
Hen erythrocyte	○	○
Cell combination in heterokaryon		
HeLa–HeLa	+ +	+ +
HeLa–rabbit macrophage	+ +	+ +
HeLa–rat lymphocyte	+ +	+ +
HeLa–hen erythrocyte	+ +	+ +
Rabbit macrophage–rabbit macrophage	+ +	○ ○
Rabbit macrophage–rat lymphocyte	+ +	○ ○
Rabbit macrophage–hen erythrocyte	+ +	○ ○

○, No synthesis in any nuclei; ○ ○, no synthesis in any nuclei of either type; +, synthesis in some or all nuclei; + +, synthesis in some or all nuclei of both types.

parent cells normally synthesizes DNA, then DNA synthesis will take place in both types of nuclei in the heterokaryon; (3) if neither of the parent cells normally synthesizes DNA, then no synthesis of DNA takes place in the heterokaryon. The regulation of nucleic acid synthesis in the heterokaryon is thus essentially unilateral: whenever a cell that synthesizes a particular nucleic acid is fused with one that does not, the active cell initiates this synthesis in the inactive partner. The inactive cell does not suppress synthesis in the active partner, even in those heterokaryons in which a number of inactive cells are fused with one active cell.

The reactivation of the erythrocyte nucleus

Two special features make heterokaryons in which one of the parent cells is a nucleated erythrocyte especially suitable for the study of nuclear reactivation. The first is that it is possible to introduce the erythrocyte nucleus into another cell either with or without the erythrocyte cytoplasm. 'Sendai' virus is a haemolytic virus, and when adult hen erythrocytes are treated with the concentrations of virus required for the formation of heterokaryons, the erythrocytes undergo complete haemolysis. The erythrocyte nuclei, essentially free of their own cytoplasm, are then introduced into the cytoplasm of the recipient cells by the fusion of these cells with the erythrocyte ghosts.[10] Erythrocytes taken from early chick embryos, however, are much more resistant to haemolysis and fuse with other cells in the normal way, thus contributing both nucleus and cytoplasm to the hybrid cell.[11] A second special feature of these heterokaryons is that the nucleus of one of the parent cells is, as I have described, completely dormant; mature erythrocytes thus provide nuclei that are particularly favourable material for the study of the mechanisms that govern the transcription and replication of DNA.

The very fact that interspecific hybrid cells may function in a perfectly integrated way, at once tells us something of importance about these mechanisms. We can be confident that the signals that the hybrid cytoplasm transmits to the genes of one of the species in the hybrid cell do not represent false signals to the genes of the other species. If such false signals were given, the end result would be a progressive disorganization of cell

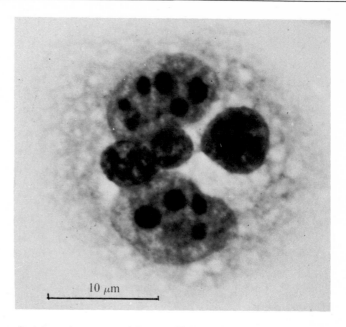

10 μm

FIG. 2.8. A heterokaryon containing two HeLa nuclei and three frog erythrocyte nuclei at various stages of reactivation.

metabolism. But we know that some interspecific hybrid cells actually multiply more vigorously than either of their parent cells; so we can dismiss the idea that the signals emanating from the hybrid cytoplasm are misunderstood by either set of genes. And this must mean that each set of genes reacts only to signals from its own cytoplasmic components, or that the signals that the hybrid cytoplasm transmits to the genes produce the same effect on both sets of genes. The demonstration that erythrocyte nuclei can be introduced into other cells without any appreciable contribution of erythrocyte cytoplasm[10] permits us to decide which of these two alternatives is correct. The experiments I have already described make it clear that reactivation of the hen erythrocyte nucleus does not require the activity of hen cytoplasm. Erythrocyte nuclei, freed of their own cytoplasm by the haemolytic action of the 'Sendai' virus, can be reactivated in the cytoplasm of cells from a wide variety of animal species, ranging from mouse to man; and frog erythrocyte nuclei can also be reactivated in these cells[7] (Fig. 2.8). It is

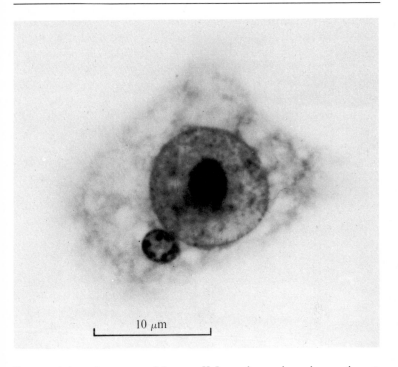

Fig. 2.9. A heterokaryon containing one HeLa nucleus and one hen erythrocyte nucleus. 'Nuclear bodies', which are highly condensed regions of chromatin, are seen in the erythrocyte nucleus.

thus obvious that these nuclei *do* respond to signals emanating from grossly foreign cytoplasm. The remarkable integration of interspecific hybrid cells cannot therefore be due to the fact that each set of genes responds only to signals from its own cytoplasmic components. The reactivation of hen erythrocyte nuclei in human or mouse cytoplasm involves not only the resumption of nucleic acid synthesis, but also, as I shall show later, the ordered synthesis of specific proteins determined by these nuclei. We can therefore conclude that the signals emanating from human or mouse cytoplasm are understood perfectly well by hen nuclei. In short, these cytoplasmic signals are not species-specific.

The outstanding morphological event associated with the reactivation of the erythrocyte nucleus is a massive increase in volume.[12] While accurate measurements of nuclear volume are

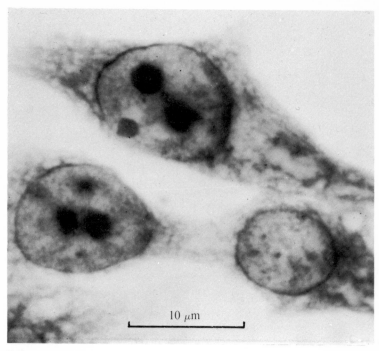

Fig. 2.10. A heterokaryon containing one HeLa nucleus and one hen erythrocyte nucleus which has been reactivated. The erythrocyte nucleus is now greatly enlarged and the chromatin within it is much less condensed.

difficult, it is likely that there is at least a 20- to 30-fold increase.[13] This expansion of the nucleus is accompanied by dispersion of its highly condensed chromatin,[12] a process that is easily demonstrated with appropriate cytological stains (Figs. 2.9, 2.10). When RNA synthesis is measured by autoradiographic means in individual enlarged erythrocyte nuclei, it is found that there is a direct relationship between the volume of the nucleus and the amount of RNA synthesized within it (Fig. 2.11): the bigger the nucleus, the more RNA it makes.[12] The increase in volume of the nucleus is not simply due to the ingress of water; there is at least a 4- to 6-fold increase in dry mass, which is largely accounted for by an increase in protein content.[13] If the erythrocytes are irradiated with a large dose of ultraviolet light before the heterokaryons are made, synthesis of RNA and DNA in the erythrocyte nuclei is largely sup-

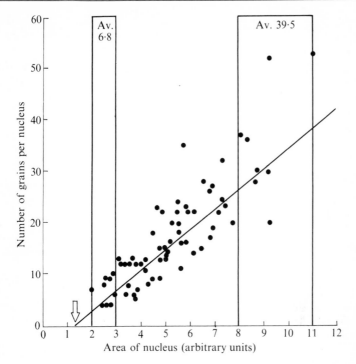

FIG. 2.11. The relationship between the maximum cross-sectional area of erythrocyte nuclei (a measure of their volume) in heterokaryons and the number of grains overlying these nuclei in autoradiographs. The cells were exposed for 20 min to a radioactive RNA precursor. The arrow indicates the mean cross-sectional area of unenlarged erythrocyte nuclei.

pressed; but the irradiated nuclei nevertheless undergo enlargement in the usual way.[12] This means that the increase in volume that the erythrocyte nuclei undergo on reactivation is not the consequence of the increased synthesis and accumulation of RNA and DNA; enlargement is the cardinal event and the progressive increase in the synthesis of nucleic acid is governed by it. When they enlarge in the heterokaryon, erythrocyte nuclei in which RNA synthesis has been suppressed by ultraviolet irradiation show the same increase in dry mass as unirradiated nuclei.[14] The irradiated nuclei cannot be synthesizing their own proteins under these conditions, so that the increase in dry mass which the erythrocyte nuclei undergo on reactivation must be due very largely to a flow of proteins from the

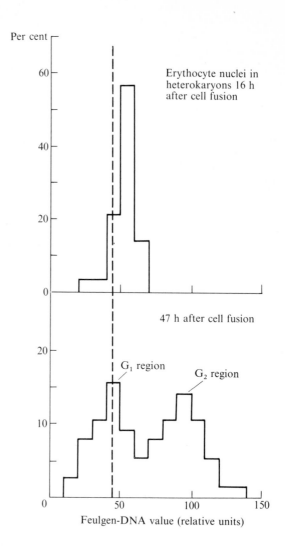

FIG. 2.12. Histogram showing the Feulgen-DNA values of erythrocyte nuclei in hetero-karyons 16 and 47 h after cell fusion. The dotted line indicates the Feulgen value of nuclei isolated from adult hen erythrocytes. It will be seen that, within 2 days, many of the erythrocyte nuclei in the heterokaryons have replicated their DNA completely.

cytoplasm into these nuclei. These proteins are, of course, human or mouse proteins, as the case may be, but they are nevertheless able to do whatever they need to do in the hen nucleus. Until the hen nucleus determines the synthesis of hen proteins in the heterokaryon, we must regard it as operating, and operating perfectly well, in an environment composed for the most part of foreign proteins.

The synthesis of DNA in the erythrocyte nuclei is not so closely linked to the process of nuclear enlargement as the synthesis of RNA; but some degree of enlargement must take place before DNA synthesis can be initiated.[15] Once initiated, however, DNA synthesis in the two sets of nuclei in the heterokaryon is rapidly and effectively synchronized, especially in binucleate cells.[15] Microspectrophotometric measurements show that the amount of DNA in the erythrocyte nuclei increases progressively and eventually reaches the tetraploid value, thus showing that the genetic material is replicated completely (Fig. 2.12).[13] These findings indicate that the replication of DNA in the hen nucleus, as well as its transcription, can be accurately controlled by signals emanating from human or mouse cytoplasm.

The dispersion of the chromatin which takes place as the nuclei enlarge is associated with certain structural changes that can be measured by cytochemical techniques. Of these changes, the most striking is a markedly increased affinity of the chromatin for intercalating dyes such as acridine orange and ethidium bromide.[13] Even before replication of DNA begins, there is at least a four-fold increase in the amount of acridine orange which the chromatin can bind (Fig. 2.13), and the amount of dye bound increases still further as nuclear enlargement proceeds. When the ability of the chromatin to bind actinomycin D is measured by an assay that defines the specific 'tight' binding between the antibiotic and the DNA, a similar increase is observed as the nucleus enlarges.[16] Reactivation of the chromatin is also accompanied by changes in its melting profile which reveal an increased susceptibility to heat denaturation (Fig. 2.14).[13] All these observations reflect the fact that nuclear enlargement loosens the chromatin and renders it more accessible, not only to macromolecules, but even to smaller molecules such as the acridine dyes and actinomycin. The same process no doubt also renders the chromatin more accessible to the molecules involved in its transcription, so that, as more of the initially condensed chromatin opens up, more of it is transcribed.

Many of the cytochemical changes that the erythrocyte

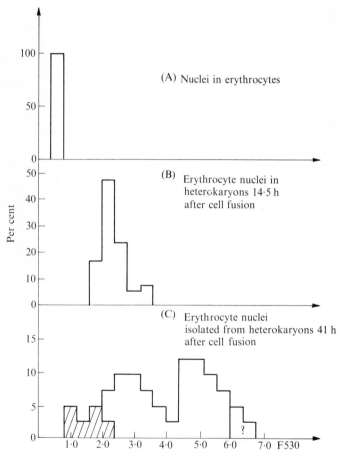

FIG. 2.13. The acridine-orange fluorescence at 530 nm of nuclei in intact erythrocytes (A), G1 phase erythrocyte nuclei in heterokaryons 14·5 h after cell fusion (B), and erythrocyte nuclei isolated from heterokaryons 41 h after cell fusion (C). The reactivation of the erythrocyte nuclei is associated with a great increase in the binding of the acridine orange to DNA. (The cross-hatched area indicates a group of small, very dense nuclei, and the question mark a group of nuclei which could not with certainty be identified as erythrocyte nuclei.)

nucleus undergoes when it is reactivated in the heterokaryon can be mimicked in erythrocyte ghosts if these are treated under certain conditions with agents which chelate divalent cations.[13,14] It is therefore not improbable that the mechanisms by which the transcription of DNA is regulated in these nuclei may involve, among other things, the interaction of the chromatin with divalent cations and possibly other electro-

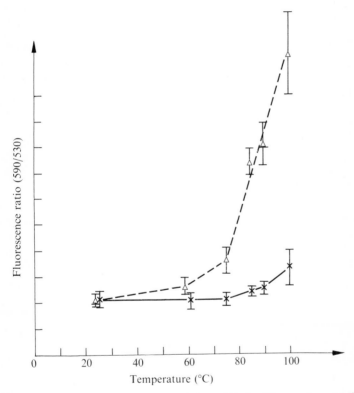

FIG. 2.14. The melting profile of the chromatin in nuclei isolated from erythrocytes (X) and in erythrocyte nuclei isolated from heterokaryons 46 h after cell fusion (△). The melting profiles were determined by measuring the fluorescence of acridine orange bound to the DNA, at 590 nm and at 530 nm. The green fluorescence (590 nm) measures the amount of dye bound to double-stranded DNA, the red fluorescence (530 nm) measures the amount bound to single-stranded DNA. The ratio of the fluorescence values at the two wavelengths indicates the proportion of the DNA which has undergone denaturation at any given temperature. It is clear that reactivation of the erythrocyte nuclei renders their chromatin much more susceptible to denaturation by heat.

lytes. Specificity of regulation, that is, the transcription of some genetic areas and not others, would then reside, not in the specificity of the cytoplasmic signals, but in the structure of the chromatin itself, which would determine, for any particular electrolyte environment, which genetic areas would be condensed, and hence inactive, and which would be opened up and hence active. In any case, if the reactivation of the dormant erythrocyte nuclei is achieved, at least in part, by changes in the immediate electrolyte environment of the chromatin, it is not surprising that this reactivation takes place in cytoplasm from widely different animal species. On this view, the signals that pass to the hen nuclei from human or mouse cytoplasm would be of a quite general kind likely to be common to all vertebrate cells.

Persistence or loss of phenotypic characters

While it is clear that, when the two are fused together, a cell that synthesizes RNA or DNA will induce this synthesis in a cell that does not, the analysis of the expression of specific phenotypic traits does not yield so clear a result. Ephrussi and his colleagues,[17-20] using Littlefield's technique[21] to select for hybrids that result from the spontaneous fusion of cells in mixed cultures, have studied the formation of pigment and the expression of certain forms of morphological differentiation in hybrid cells in which one of the parents showed the particular character being investigated and the other did not. When pigmented cells derived from a melanotic tumour of the Syrian hamster were fused with unpigmented cells from three different mouse cell lines, the hybrids that were examined failed to produce pigment or dopa oxidase or tyrosinase, enzymes essential for pigment synthesis.[17-19,22] When cells derived from a mouse testicular teratocarcinoma, and still capable of various forms of specific morphological differentiation, were fused with cells from a fibroblastic mouse cell line, the hybrids examined were fibroblastic in character and did not appear to retain the ability to undergo the specific forms of morphological differentiation of which the teratocarcinoma cells were capable.[20] These findings have been interpreted as evidence for the view that certain forms of differentiation are under 'negative control',[18] an unilluminating phrase which, in the absence of any precise understanding of the mechanisms that regulate the synthesis of the relevant enzymes, does little more than re-phrase the observed fact that the particular differentiated trait under investigation does not happen to appear in the hybrid cell. The interpretation of these experiments is difficult. The differentiated cells chosen for study were highly unusual cell types, in that, in both cases, they were derived from malignant tumours and were the end-products of complicated procedures designed to select for cells that either retained the ability to express the differentiated state in vitro[23,17] or retained the ability to express the differentiated state when re-injected into the animal.[20,24,25] It is, of course, a commonplace experi-ence that differentiated cells rapidly lose their differentiated character when cultivated in vitro; and it is only very recently that conditions have been found which permit the growth of highly differentiated cells in artificial culture.[26-28] All stable lines of differentiated cells that have so far been produced appear to have been derived from malignant tumours by pro-cedures involving prolonged selection.[23-25,29] We are therefore dealing in these cases, not only with cells showing a particular form of differentiation, but also with malignant cells which have been selected for their ability to maintain certain features of the differentiated state on prolonged cultivation in vitro. What is abolished when such cells are fused with other cells is not only the differentiated state itself, but also the specialized organization that permits these highly selected cells to maintain the differentiated state in vitro. The situation is further compli-cated by the fact that in some cell types, specialized function in vitro is abolished when the differentiated cells are exposed to the antimetabolites normally used to select for hybrid cells in Littlefield's procedure;[30] and some differentiated cells lose their differentiated character even when they are grown in contact with undifferentiated cells.[31] It is certainly of interest that fusion of these specialized cells with some other cells abolishes the or-ganization that determines the maintenance of a particular dif-ferentiated state in vitro; but it is far from clear that this result permits any useful generalization about the 'positive' or 'nega-tive' nature of the controls that govern differentiation as a whole. Indeed, when hepatic cells, derived directly from em-bryonic mouse liver, were fused with mouse L cells, clones of hybrid cells were formed which showed the morphological features, and apparently also some of the biochemical features, of the hepatic cells;[32] and when mouse cells producing a high level of interferon were fused with hamster cells producing a low level, the hybrid cells appeared to produce not only a high level of mouse interferon, but also about ten times as much hamster interferon as the parent hamster cell.[33] These, then, could be interpreted as examples of 'positive' control. But when mouse cells producing low levels of collagen and hyaluronic acid were fused with others producing high levels of these substances, the hybrid cells were found to produce intermediate amounts of both substances,[34] thus revealing neither 'positive' nor 'negative' control, or, alternatively, a balanced mixture of both. A similar result was obtained in hybrids between cells with high and low folate reductase activity.[35] I think enough experiments of this type have already been done to make it improbable that any simple general rule will adequately describe the processes that control differentiation. This is hardly surprising. Even in the much simpler bacterial systems that have been examined, enzyme synthesis in the heterozygote may show either 'negative'[36] or 'positive'[37] control, or more complicated forms of regulation that cannot easily be described as either 'negative' or 'positive'.[38,39]

Drastic changes in phenotype, and especially morphological changes, are often produced immediately, or very shortly, after cell fusion, and probably do not depend on interaction at the genetic level. For example, when highly specialized cells such as macroblages or lymphocytes are fused with HeLa cells or with fibroblasts, many of the specialized features of the leucocytes are very rapidly eliminated.[2] Macrophages and lymphocytes are amoeboid cells, each with a characteristic mode of locomotion. The macrophage has a very pronounced peripheral 'undulating membrane', is highly phagocytic, and exhibits chemotaxis. The lymphocyte has a surface which does not adhere to glass. The macrophage has the ability to ingest erythrocytes coated with antibody; and its membrane contains a specific adenosine triphosphatase.[40] All these properties of the leucocytes disappear within a few hours when these cells are fused with HeLa or other tissue culture cells, even when several macrophages or lymphocytes are fused with one of the undifferentiated cells;[2,40] and, in the case of the macrophage, it has been shown that this disappearance of specific surface properties does not require RNA synthesis by the macrophage nucleus.[41] Nor are any of these differentiated traits restored in the progeny of the original fused cell. When mitosis and nuclear fusión occur, the resulting synkaryon still shows no features easily recognizable as leucocytic,[2] and hybrid cell lines derived from such synkaryons still retain an essentially undifferentiated character.[42,43] These effects are not happily described in terms of genetic 'dominance' or 'recessiveness'. The characteristic behaviour of these highly differentiated cells depends upon the establishment, during the process of differentiation, of specialized forms of cytoplasmic organization, often structural organization. This organization may be disrupted by the act of fusion itself, or may break down shortly afterwards; and it is rather sanguine to expect that it would be re-established in the hybrid cell by the process of cultivation *in vitro*. If one wished to regenerate in the hybrid the specialized function of the differentiated parent cell, one might perhaps do better to expose the hybrid cells to the environmental influences that generated the differentiated state in the first instance. When undifferentiated tumour cells are fused with myoblasts undergoing differentiation to form muscle syncytia *in vitro*, the tumour cells do not inhibit the differentiation, but are incorporated into the myotubes and form part of the muscle tissue.[43]

It sometimes happens that the hybrid cell shows features that are not present in either of the parent cells. The cells of the Ehrlich ascites tumour of the mouse are roughly spherical objects which do not adhere to glass; HeLa cells grow well on glass in typical epithelial sheets. But the hybrid cells derived from the fusion of these two cells have an irregular fibroblastic morphology on glass and are unusually motile.[44,45] The macrophage adheres to glass and has a characteristic amoeboid motion; but when it is fused with the spherical, non-adherent Ehrlich ascites cell, the hybrid has a stellate morphology on glass and shows long slender cytoplasmic processes.[46] Multinucleate cells in which the macrophage is the dominant partner often distribute their nuclei around the periphery of the cell[2] (Figs. 2.6, 2.7), an arrangement well described by Langhans[47,48] in the last century. These new morphological features in the fused cell no doubt represent attempts to solve the problem of marrying different, and perhaps antagonistic, forms of cytoplasmic organization. Sometimes the solution is successful, and the new morphological feature is inherited in the progeny of the original hybrid cell; sometimes the solution is unsuccessful, and the hybrid is not viable.

Another word of caution to discourage facile generalization. When cells of two different types are fused together, the properties of the resultant hybrids are not always the same. When Ehrlich ascites cells were fused with mouse cells of a fibroblastic type, and the individual hybrid clones arising from each act of fusion were examined, great variation was found from clone to clone in cell shape, clonal morphology, and growth rate.[49] Tight epithelial clones, clones composed of long spindle-shaped cells growing in parallel, highly dispersed clones of irregular fibroblastic cells, and a whole range of intermediate categories were observed. It is therefore essential, before one can attempt generalizations about the persistence or loss of any phenotypic character in a hybrid cell, that an adequate number of clones derived from primary fusions be examined. This is especially the case where the progeny of rare spontaneous fusions are obtained by the use of selective media, or where the hybrid cells overgrow their parents in bulk culture. In the first case, the hybrids one obtains may be atypical; in the second, one simply selects those hybrids which grow most rapidly.

Regulation of haemoglobin synthesis

One highly differentiated function which has been analysed in some detail by means of cell fusion is the synthesis of haemoglobin.[11] As mentioned earlier, chick erythrocyte nuclei may be introduced into the cytoplasm of other cells either with or without erythrocyte cytoplasm. In adult hen erythrocytes, or erythrocytes taken from 12- to 15-day-old chick embryos, the inactivated 'Sendai' virus induced haemolysis which releases the cytoplasmic contents of the erythrocytes before cell fusion is achieved.[10] In these mature erythrocytes, the synthesis of haemoglobin has very largely come to an end. In very immature erythrocytes, taken from 3- to 5-day-old chick embryos, 'Sendai' virus does not induce haemolysis, and cell fusion takes place in the usual way. These immature erythrocytes synthesize haemoglobin, so that cell fusion introduces into the recipient cell not only the erythrocyte nucleus, but also erythrocyte cytoplasm actually engaged in haemoglobin synthesis (Fig. 2.15).[11] When erythrocyte nuclei alone are introduced into tissue culture cells of the same, or different, species, haemoglobin synthesis cannot be detected in the heterokaryons, even though the erythrocyte nuclei, as I shall show in the next chapter, do determine the synthesis of other chick proteins. It thus appears that the cytoplasm of the recipient cell does not induce the erythrocyte nucleus to synthesize templates for haemoglobin synthesis; or, if it does induce the synthesis of these templates, it does not permit their translation into protein. This is hardly surprising. For where the recipient cell is a normal diploid cell, for example, a chick fibroblast, we must assume that its nucleus already contains the genes that determine the synthesis of haemoglobin; and since the fibroblast does not synthesize haemoglobin despite the fact that its nucleus contains the necessary genes, it is difficult to see why haemoglobin synthesis should be induced in the fibroblast cytoplasm simply by the introduction of a second set of the same genes.

However, if cytoplasm actually engaged in haemoglobin synthesis is introduced into another cell, a very different result is obtained. When early embryonic erythrocytes, still synthesizing haemoglobin, are fused with tissue culture cells, haemoglobin synthesis is at first markedly stimulated: during the first few hours after cell fusion, the rate of haemoglobin synthesis in the heterokaryons may rise to levels ten times as high as those found in the erythrocytes themselves. But within the first day after fusion, the rate of haemoglobin synthesis begins to fall, and synthesis is completely abolished by the fourth or fifth day (Fig. 2.16).[11] Neither the stimulation nor the suppression of haemoglobin synthesis in the heterokaryon is yet understood. It is possible that the initial stimulation is due to the fact that the general metabolic level in a multiplying tissue culture cell is much higher than that in a differentiating erythrocyte, so that, once templates for the synthesis of haemoglobin are available in the heterokaryon, translation of these templates proceeds at a faster rate. This view is perhaps supported by the observation that, if the cells are subjected to excessive trauma during the process of cell fusion, as occurs, for example, when

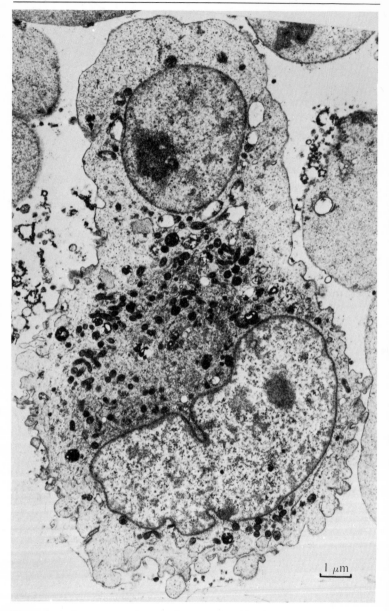

1 μm

FIG. 2.15. An immature erythrocyte, from a 5-day-old chick embryo, in the process of fusing with a mouse tissue culture cell. Erythrocyte cytoplasm engaged in haemoglobin synthesis is thus introduced into the hybrid cell.

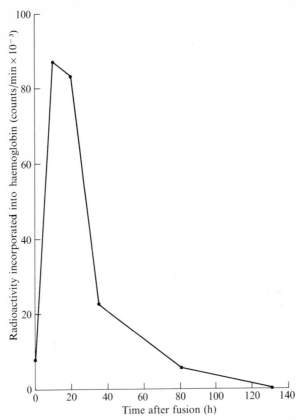

FIG. 2.16. Synthesis of haemoglobin in heterokaryons produced by fusing immature erythrocytes with mouse tissue culture cells. There is initially a great increase in the rate of haemoglobin synthesis followed by a gradual decrease. The cultures contained about 120 erythrocyte nuclei per 100 cells.

very high doses of inactivated virus are used, the stimulation of haemoglobin synthesis may be much reduced, or may not occur at all. It is also possible that synthesis of haemoglobin in the erythrocyte is limited by the availability of haem[50] and that the tissue culture cell, which synthesizes haem for the production of cytochrome B and certain enzymes, may make more haem available. There is as yet no experimental information about the mechanism by which the more gradual, but permanent, suppression of haemoglobin synthesis is achieved. The possibility that this process may involve the destruction of the RNA derived from the erythrocyte is being explored. It is, in any case, clear that haemoglobin synthesis may be both stimulated and suppressed by the cytoplasm of the recipient cell, and the evidence, so far, appears to indicate that this regulation does not operate through the erythrocyte nucleus.

Histocompatibility antigens

The study of the expression of histocompatibility antigens provides an interesting illustration of the complexities involved in the analysis of 'dominance' and 'recessiveness' of phenotypic traits in hybrid cells. The first investigation of this problem dealt with the expression of histocompatibility antigens in hybrid cells arising spontaneously in mixed cultures of three mouse cell lines which differed in the antigens determined by the H-2 locus.[51] In four clones from one cross between two of these cell lines, and in one clone from another cross, it was found that the histocompatibility antigens of both parent cells were expressed in the hybrids. A similar result was also obtained for two surface antigens not determined by the H-2 locus. It was concluded that the expression of histocompatibility antigens was 'co-dominant'. As a restatement of the observed facts in these particular hybrids, 'co-dominance' is no doubt a perfectly acceptable term; but the authors of these experiments argue that 'dominance' is a general characteristic of histocompatibility antigens, and, indeed, that these antigens must be essential for cell survival. A recent re-examination of this question shows that the position is, in fact, more complicated than this.[52] It was confirmed that hybrids between cells having the same or different antigens determined by the H-2 locus did, in general, show the antigens of both parent cells; but, in some cases, there were quantitative differences between the parent cells and the hybrids. Moreover, when Ehrlich ascites cells, which grow in any strain of mouse, were fused with mouse cells bearing the $H-2^k$ antigen complex, the hybrid cells that overgrew the mixed cultures showed very little $H-2^k$ antigen when tested by immune fluorescence and immune adherence tests. When these hybrids were tested by a more sensitive mixed haemadsorption test, small amounts of the $H-2^k$ antigen could be detected on some of the hybrid cells, but many failed to show these antigens. The chromosome constitution of these hybrids was approximately what one would expect from the fusion of the two parent cells. One might therefore be tempted to conclude that, with this particular combination of cells, the expression of the $H-2^k$ antigens was 'recessive'. However, when clones derived from a number of different primary fusions were examined, it was found that the degree of expression of the $H-2^k$ locus varied: in some clones these antigens could barely be detected, or could not be detected at all; in others they were clearly present. It thus appears that the expression of histocompatibility genes in hybrid cells (as in ordinary diploid cells in the body) is subject to important quantitative variations, which may result in apparent 'dominance' in one type of hybrid and apparent 'recessiveness' in another.

Malignancy

Cell fusion has also been used to study the dominance relationships of malignancy. The original mouse cell hybrid described by Barski, Sorieul, and Cornefert[53] arose from the fusion of cells derived from two lines which differed in their ability to produce tumours: one of the parent cells was highly malignant, the other less so. Fifteen separate clones of hybrid cells derived from these two parents were examined and, in fourteen of these, the ability of the hybrids to produce tumours more closely resembled that of the more malignant parent.[54] Hybrids between the more malignant of these two parent cells and normal mouse fibroblasts also appeared to be malignant.[55] The conclusion was drawn that malignancy was a 'dominant' character. An essentially similar conclusion was drawn from experiments on some spontaneous hybrids arising in mixed cultures of nonmalignant mouse cells and mouse cells that had been rendered malignant by infection with polyoma virus. The hybrids examined again appeared to be malignant and, in some cases, showed an even greater ability to produce tumours than the malignant parent cell.[56] In this case, it could also be shown that antigens determined by the polyoma virus continued to be produced in the hybrid cells.[56,57] However, it is by no means a general rule that tumour viruses persist in hybrid cells or that malignancy is invariably dominant. A more recent study has again revealed a situation of greater complexity. When the highly malignant cells of the Ehrlich ascites tumour were fused with cells from a mouse cell line of very low malignancy, a number of hybrid cell lines were derived that contained both parental chromosomal sets but were not malignant, even when assayed (both intraperitoneally and subcutaneously) in X-irradiated, new-born mice of the appropriate histocompatibility group.[52] Moreover, these hybrid cells did not show the highly aberrant cell cycle characteristic of Ehrlich ascites cells.[58] In logarithmic growth, Ehrlich ascites cells have virtually no G_1 phase and synthesize DNA continuously throughout most of interphase;[59] but the hybrid cells appeared to have a normal cell cycle with a normal G_1 phase. It thus appears that the Ehrlich cell can be cured both of its malignancy and of its abnormal cell cycle by fusion with a cell that is not malignant and has a normal cell cycle. One of the cells studied in this experiment bore an antigen characteristic of the Moloney type of tumour virus; the other cell had no such antigen. Although the amount of this antigen in the hybrid cells varied, some hybrid clones were found in which the expression of this antigen was suppressed.[52] It therefore appears that both malignancy and the expression of a tumour antigen can be suppressed by cell fusion.

This suppression of malignancy has been analysed further by an examination of segregants derived from the hybrid cells by selection *in vivo*.[60] Although the vast majority of these hybrid cells were not malignant, occasional animals into which large numbers of cells had been injected intraperitoneally developed ascites tumours several weeks or months after injection. These secondary tumours could be transplanted serially to other animals. When the cells of the tumours were cultivated *in vitro*, their behaviour was found to be very different from that of the original hybrid cells or that of either of the parent cells. Nevertheless, these tumour cells had evolved from the hybrid cells injected into the animal, for they were found to contain marker chromosomes derived from both of the original parent cells. The average number of chromosomes in the tumour cells was, however, much lower than that of the hybrid cells injected: some 40 chromosomes out of a total of nearly 130 had been lost.[61] The cells of the secondary tumours were thus segregants of the original hybrids, which had reverted to malignancy.

These experiments thus appeared to show that when the non-malignant cell and the malignant cell were fused together, the non-malignant cell contributed something to the hybrid, which suppressed the malignancy. This contribution was apparently linked to the activity of one or more of the chromosomes of the non-malignant cell, since the elimination of chromosomes from the hybrid resulted in a reversion to malignancy. As all malignant segregants can be selected for by growth *in vivo*, a more detailed analysis of the phenomenon will clearly be possible. That malignancy can be suppressed in this way and that the suppression is amenable to genetic analysis are findings that can hardly be irrelevant to the future progress of cancer research.

Cell fusion has also thrown some new light on another important problem in this field. It is well known that certain tumour viruses may produce transformations in susceptible cells which endow them with the property of malignancy. This induced malignancy is heritable and the transformed cells continue to synthesize antigens characteristic of the tumour viruses to which they have been exposed. However, these cells may altogether lose the ability to produce infective virus; and some transformed cell lines will not yield infective virus under any of the conditions that normally do elicit the production of such viruses. But, in the case of SV40 virus and Rous sarcoma virus, cells that have been transformed, but do not yield any infective virus, can be induced to do so by fusion with susceptible normal cells.[62-66] In some of the heterokaryons formed in this way, infective virus is again produced, and the resuscitated virus is apparently genetically identical with the virus that produced the original transformation.[67] It thus appears that the cell can harbour tumour viruses in a latent or masked form, but such viruses may be unmasked by appropriate cell fusion. In contemporary terms, this presumably means that, when a virus is being carried in a latent form, some of its genes fail to be expressed; fusion with another cell may induce their expression. This discovery has initiated a search for other latent tumour viruses in man as well as in laboratory animals, and, like the discovery of lysogeny in bacteria, it has opened the possibility of analysing the behaviour of such viruses in new ways.[68-72]

Application of cell fusion to problems in mammalian embryology

Finally, I should like to mention a new application of cell fusion which may prove to be of great importance. The 'Sendai' virus technique has recently been successfully applied to mammalian ova and to cells from early stages of embryological development. Ova can be fused with each other, and the nucleus of a completely differentiated adult cell can be introduced into the ovum. These developments are the beginnings of a new approach to the analysis of early embryological development,[73] and they open the possibility of generating new mammals from single somatic cells, a feat that has hitherto been possible only with the large extracorporeal eggs of amphibia.[74,75] The significance of these advances does not need to be elaborated.

REFERENCES

1. HARRIS, H. (1965). Behaviour of differentiated nuclei in heterokaryons of animal cells from different species. *Nature, Lond.* **206**, 583.
2. HARRIS, H., WATKINS, J. F., FORD, C. E., and SCHOEFL, G. I. (1966). Artificial heterokaryons of animal cells from different species. *J. Cell Sci.* **1**, 1.
3. MACKANESS, G. B. (1952). The action of drugs on intracellular tubercle bacilli. *J. Path. Bact.* **64**, 429.
4. WATTS, J. W. and HARRIS, H. (1959). Turnover of nucleic acids in a non-multiplying animal cell. *Biochem. J.* **72**, 147.
5. GOWANS, J. L. and KNIGHT, E. J. (1964). The route of recirculation of lymphocytes in the rat. *Proc. R. Soc.* B**159**, 257.
6. GOWANS, J. L., MacGREGOR, D. D., COWEN, D. M., and FORD, C. E. (1962). Initiation of immune responses by small lymphocytes. *Nature, Lond.* **196**, 651.
7. HARRIS, H. (1966). Hybrid cells from mouse and man: a study in genetic regulation. *Proc. R. Soc.* B**166**, 358.
8. HARRIS, H., SIDEBOTTOM, E., GRACE, D. M., and BRAMWELL, M. E. (1969). The expression of genetic information: a study with hybrid animal cells. *J. Cell Sci.* **4**, 499.
9. HARRIS, H. Unpublished results.
10. SCHNEEBERGER, E. E. and HARRIS, H. (1966). An ultrastructural study of interspecific cell fusion induced by inactivated Sendai virus. *J. Cell Sci.* **1**, 401.
11. COOK, P. R. and HARRIS, H. In preparation.

12. HARRIS, H. (1967). The reactivation of the red cell nucleus. *J. Cell Sci.* **2**, 23.

13. BOLUND, L., RINGERTZ, N. R., and HARRIS, H. (1969). Changes in the cytochemical properties of erythrocyte nuclei reactivated by cell fusion. *J. Cell Sci.* **4**, 71.

14. RINGERTZ, N. R. and BOLUND, L. (1969). 'Activation' of hen erythrocyte deoxyribonucleoprotein. *Expl Cell Res.* **55**, 205.

15. JOHNSON, R. T. and HARRIS, H. (1969). DNA synthesis and mitosis in fused cells. 2. HeLa-chick erythrocyte heterokaryons. *J. Cell Sci.* In press.

16. RINGERTZ, N. R. Personal communication.

17. DAVIDSON, R. L., EPHRUSSI, B., and YAMAMOTO, K. (1966). Regulation of pigment synthesis in mammalian cells, as studied by somatic hybridization. *Proc. natn. Acad. Sci. U.S.A.* **56**, 1437.

18. DAVIDSON, R., EPHRUSSI, B., and YAMAMOTO, K. (1968). Regulation of melanin synthesis in mammalian cells, as studied by somatic hybridization. I Evidence for negative control. *J. cell. Physiol.* **72**, 115.

19. DAVIDSON, R. L. and YAMAMOTO, K. (1968). Regulation of melanin synthesis in mammalian cells, as studied by somatic hybridization. II The level of regulation of dopa oxidase. *Proc. natn. Acad. Sci. U.S.A.* **60**, 894.

20. FINCH, B. W. and EPHRUSSI, B. (1967). Retention of multiple developmental potentialities by cells of a mouse testicular teratocarcinoma during prolonged culture *in vitro* and their extinction upon hybridization with cells of permanent lines. *Proc. natn. Acad. Sci. U.S.A.* **57**, 615.

21. LITTLEFIELD, J. W. (1964). Selection of hybrids from matings of fibroblasts *in vitro* and their presumed recombinants. *Science, N.Y.* **145**, 709.

22. SILAGI, S. (1967). Hybridization of a malignant melanoma cell line with L-cells *in vitro*. *Cancer Res.* **27**, 1953.

23. MOORE, G. E. (1964). *In vitro* cultures of a pigmented hamster melanoma cell line. *Expl Cell Res.* **36**, 422.

24. STEVENS, L. C. (1958). Studies on transplantable testicular teratomas of strain 129 mice. *J. natn. Cancer Inst.* **20**, 1257.

25. STEVENS, L. C. (1960). Embryonic potency of embryoid bodies derived from a transplantable testicular teratoma of the mouse. *Devl Biol.* **2**, 285.

26. CAHN, R. D. and CAHN, M. B. (1966). Heritability of cellular differentiation: clonal growth and expression of differentiation in retinal pigment cells *in vitro*. *Proc. natn. Acad. Sci. U.S.A.* **55**, 106.

27. COON, H. G. (1966). Clonal stability and phenotypic expression of chick cartilage cells *in vitro*. *Proc. natn. Acad. Sci. U.S.A.* **55**, 66.

28. YAFFE, D. (1968). Retention of differentiation potentialities during prolonged cultivation of myogenic cells. *Proc. natn. Acad. Sci. U.S.A.* **61**, 477.

29. YASUMURA, Y., TASHJIAN, A. H., and SATO, G. H. (1966). Establishment of four functional, clonal strains of animal cells in culture. *Science, N.Y.* **154**, 1186.

30. LASHER, R. and CAHN, R. D. (1969). The effects of 5-bromodeoxyuridine on the differentiation of chondrocytes *in vitro*. *Devl Biol.* **19**, 415.

31. BRYAN, J. (1968). Studies on clonal cartilage strains. I Effect of contaminant non-cartilage cells. *Expl Cell Res.* **52**, 319.

32. COON, H. G. (1969). Quantitative studies of Sendai assisted somatic

hybrid strain formation between two 'L' cell strains, and 'L' cells and mammalian liver cells. *Proceedings of the Wistar Institute Symposium on Heterospecific Genome Interaction.* In press.

33. CARVER, D. H., SETO, D. S. Y., and MIGEON, B. R. (1968). Interferon production and action in mouse, hamster, and somatic hybrid mouse–hamster cells. *Science, N.Y.* **160**, 558.

34. GREEN, H., EPHRUSSI, B., YOSHIDA, M., and HAMERMAN, D. (1966). Synthesis of collagen and hyaluronic acid by fibroblast hybrids. *Proc. natn. Acad. Sci. U.S.A.* **55**, 41.

35. LITTLEFIELD, J. W. (1969). Hybridization of hamster cells with high and low folate reductase activity. *Proc. natn. Acad. Sci. U.S.A.* **62**, 88.

36. JACOB, F. and MONOD, J. (1961). Genetic regulatory mechanisms in the synthesis of proteins. *J. molec. Biol.* **3**, 318.

37. ENGLESBERG, E., IRR, J., POWER, J., and LEE, N. (1965). Positive control of enzyme synthesis by gene C in the L-arabinose system. *J. Bact.* **90**, 946.

38. McFALL, E. (1967). Dominance studies with stable merodiploids in the D-serine deaminase system of *Escherichia coli* K12. *J. Bact.* **94**, 1982.

39. McFALL, E. (1967). 'Position effect' on dominance in the D-serine deaminase system of *Escherichia coli* K12. *J. Bact.* **94**, 1989.

40. GORDON, S. Personal communication.

41. GORDON, S. Personal communication.

42. MIGGIANO, V., NABHOLZ, M., and BODMER, W. (1969). Hybrids between human leucocytes and a mouse cell line: production and characterization. *Proceedings of the Wistar Institute Symposium on Heterospecific Genome Interaction.* In press.

43. DEFENDI, V. (1968). Communication to the Workshop on Virus Induction by Cell Association, held at the Wistar Institute, Philadelphia, U.S.A.

44. HARRIS, H. and WATKINS, J. F. (1965). Hybrid cells derived from mouse and man: artificial heterokaryons of mammalian cells from different species. *Nature, Lond.* **205**, 640.

45. WATKINS, J. F. and GRACE, D. M. (1967). Studies on the surface antigens of interspecific mammalian cell heterokaryons. *J. Cell Sci.* **2**, 193.

46. WATKINS, J. F. Personal communication.

47. LANGHANS, T. (1868). Ueber Riesenzellen mit wandständigen Kernen in Tuberkeln und die fibröse Form des Tuberkels. *Virchows Arch. path. Anat. Physiol.* **42**, 382.

48. LANGHANS, T. (1870). Beobachtungen über Resorption der Extravasate und Pigmentbildung in denselben. *Virchows Arch. path. Anat. Physiol.* **49**, 66.

49. HARRIS, H. Unpublished results.

50. BRUNS, G. P. and LONDON, I. H. (1965). The effect of hemin on the synthesis of globin. *Biochem. biophys. Res. Commun.* **18**, 236.

51. SPENCER, R. A., HAUSCHKA, T. S., AMOS, D. B., and EPHRUSSI, B. (1964). Co-dominance of isoantigens in somatic hybrids of murine cells grown *in vitro. J. natn. Cancer Inst.* **33**, 893.

52. HARRIS, H., MILLER, O. J., KLEIN, G., WORST, P., and TACHIBANA, T. (1969). The suppression of malignancy by cell fusion. *Nature, Lond.* **223**, 363.

53. BARSKI, G., SORIEUL, S., and CORNEFERT, F. (1960). Production dans des cultures *in vitro* de deux souches cellulaires en association, de cellules de caractère 'hybride'. *C. r. hebd. Séanc. Acad. Sci., Paris* **251**, 1825.

54. BARSKI, G. and Cornefert, F. (1962). Characteristics of 'hybrid'-type clonal cell lines obtained from mixed cultures *in vitro*. *J. natn. Cancer Inst.* **28**, 801.

55. SCALETTA, L. J. and EPHRUSSI, B. (1965). Hybridization of normal and neoplastic cells *in vitro*. *Nature, Lond.* **205**, 1169.

56. DEFENDI, V., EPHRUSSI, B., KOPROWSKI, H., and YOSHIDA, M. C. (1967). Properties of hybrids between polyoma-transformed and normal mouse cells. *Proc. natn. Acad. Sci. U.S.A.* **57**, 299.

57. DEFENDI, V., EPHRUSSI, B., and KOPROWSKI, H. (1964). Expression of polyoma-induced cellular antigen(s) in hybrid cells. *Nature, Lond.* **203**, 495.

58. MILLER, O. J. Personal communication.

59. BASERGA, R. (1963). Mitotic cycle of ascites tumor cells. *Archs Path.* **75**, 156.

60. HARRIS, H. (1969). In preparation.

61. MILLER, O. J. and HARRIS, H. (1969). In preparation.

62. GERBER, P. (1966). Studies on the transfer of subviral infectivity from SV40-induced hamster tumor cells to indicator cells. *Virology* **28**, 501.

63. WATKINS, J. F. and DULBECCO, R. (1967). Production of SV40 virus in heterokaryons of transformed and susceptible cells. *Proc. natn. Acad. Sci. U.S.A.* **58**, 1396.

64. KOPROWSKI, H., JENSEN, F. C., and STEPLEWSKI, Z. (1967). Activation of production of infectious tumour virus SV40 in heterokaryon cultures. *Proc. natn. Acad. Sci. U.S.A.* **58**, 127.

65. SVOBODA, J., MACHALA, O., and HLOŽÁNEK, I. (1967). Influence of Sendai virus on RSV formation in mixed culture of virogenic mammalian cells and chicken fibroblasts. *Folia biol., Praha* **13**, 155.

66. VIGIER, P. (1967). Persistence du génome du virus de Rous dans des cellules du hamster converties *in vitro*, et action du virus Sendai inactivé sur sa transmission aux cellules de poule. *C. r. hebd. Séanc. Acad. Sci., Paris* **264**, 422.

67. TAKEMOTO, K. K., TODARO, G. J., and HABEL, K. (1968). Recovery of SV40 virus with genetic markers of original inducing virus from SV40-transformed mouse cells. *Virology* **35**, 1.

68. KIT, S., KURIMURA, T., SALVI, M. L., and DUBBS, D. R. (1968). Activation of infectious SV40 DNA synthesis in transformed cells. *Proc. natn. Acad. Sci. U.S.A.* **60**, 1239.

69. STEPLEWSKI, Z., KNOWLES, B. B., and KOPROWSKI, H. (1967). The mechanism of internuclear transmission of SV40-induced complement fixation antigen in heterokaryocytes. *Proc. natn. Acad. Sci. U.S.A.* **59**, 769.

70. KNOWLES, B. B., JENSEN, F. C., STEPLEWSKI, Z., and KOPROWSKI, H. (1968). Rescue of infectious SV40 after fusion between different SV40-transformed cells. *Proc. natn. Acad. Sci. U.S.A.* **61**, 42.

71. BURNS, W. H. and BLACK, P. H. (1968). Analysis of Simian Virus 40-induced transformation of hamster kidney tissue *in vitro* V. Variability

of virus recovery from cell clones inducible with mitomycin C and cell fusion. *J. Virol.* **2**, 606.

72. Dubbs, D. R. and Kit, S. (1968). Isolation of defective lysogens from Simian Virus 40-transformed mouse kidney cultures. *J. Virol.* **2**, 1272.

73. Graham, C. F. (1969). The fusion of cells with one and two cell mouse embryos. *Proceedings of the Wistar Institute Symposium on Heterospecific Genome Interaction.* In press.

74. Gurdon, J. B. (1962). The developmental capacity of nuclei taken from intestinal epithelium cells of feeding tadpoles. *J. Embryol. exp. Morph.* **10**, 622.

75. Gurdon, J. B. and Uehlinger, V. (1966). 'Fertile' intestine nuclei. *Nature, Lond.* **210**, 1240.

The expression of genetic information

Analysis of genetic activity

HETEROKARYONS in which one of the parent cells is a nucleated erythrocyte present an unparalleled opportunity to examine the whole process by which genetic information is expressed in mammalian cells.[1] For purposes of analysis, the introduction of an erythrocyte nucleus into the cytoplasm of another cell is, for mammalian cells, the formal equivalent of sexual conjugation in bacteria.[2] In both cases new genetic material is inserted into the recipient cell, and in both cases the consequences of this insertion can be analysed in terms of the synthesis or nonsynthesis of new proteins. But the experiment with mammalian cells is, in some respects, more powerful. The erythrocyte nucleus is initially in a completely repressed state, and its reactivation takes place slowly enough to permit piecemeal dissection of the process. The reactivated nuclei can be reisolated from the heterokaryon at any time, and the nature of the RNA which they are making can be examined. The passage of RNA from the reactivated nuclei to the cytoplasm of the cell can be monitored; and the relationship between the transcription of genes and the synthesis of proteins specified by these genes can be determined. In short, the whole process of information transfer can be analysed in these heterokaryons with a degree of precision that is not attainable in any other biological system. In this chapter I shall present the results of one such analysis.

Surface antigen

In studying the synthesis of any protein that might be determined by the hen erythrocyte nucleus reactivated in human or mouse cytoplasm, the first requirement is to show that the protein being examined is hen, and not human or mouse, protein. It was for this reason that the first proteins chosen for investigation in these heterokaryons were species-specific surface antigens.[3] These antigens can be detected on the surface of cells in culture with great sensitivity and complete specificity by the technique of immune haemadsorption,[4,5] an application of the mixed antiglobulin reaction.[6] The principle underlying immune haemadsorption is shown in Fig. 3.1. Sensitized red cells serve

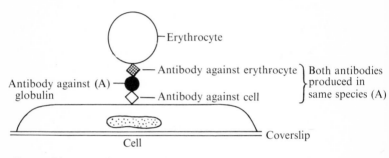

FIG. 3.1. Diagrammatic representation of the technique of mixed immune haemadsorption, a mixed antiglobulin technique. Adsorption of the sensitized erythrocyte indicates the presence of the species-specific antigen on the surface of the cell.

as the marker, and specific antiserum binds these red cells to surfaces bearing the appropriate antigens, but not to others. There is no cross-reaction, at suitable dilutions of antiserum, between hen surface antigens and human or mouse antigens, so that, in these heterokaryons, adsorbed red cells indicate the presence of hen-specific antigens. Under standard conditions, the intensity of the haemadsorption reaction also gives some indication of the amount of hen-specific antigen present.

Since the formation of these heterokaryons involves the fusion of the recipient cell with the erythrocyte ghost, membrane components derived from the erythrocyte are present in the surface of the heterokaryon immediately after fusion.[7] Immune haemadsorption reveals the presence of the hen-specific antigens (Fig. 3.2). The behaviour of these antigens during the first few days after cell fusion presents an interesting paradox. Since the hen erythrocyte nuclei undergo reactivation during this period and synthesize large amounts of RNA,[8,9] one might

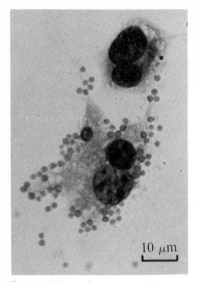

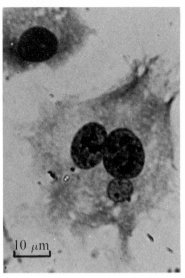

Fig. 3.2. A heterokaryon containing two mouse nuclei and one hen erythrocyte nucleus, 18 h after cell fusion. The haemadsorption reaction reveals the presence of hen-specific antigens on the surface of the cell.

Fig. 3.3. A heterokaryon containing two mouse nuclei and one hen erythrocyte nucleus, 5 days after cell fusion. The erythrocyte nucleus has been reactivated, but the absence of any haemadsorption shows that the hen-specific antigens are no longer present on the surface of the cell. The erythrocyte nucleus shows a small nucleolus.

have expected that the amount of hen-specific antigen on the surface of the heterokaryon would increase. Instead, it was found that these antigens gradually disappeared from the cell surface and, by the fourth day after fusion, could not be detected at all in the great majority of the cells (Fig. 3.3). As shown in Fig. 3.4, these antigens are retained on the surface of the cells at 20°C; and, at 28°, they are eliminated more slowly than at 37°. The hen-specific antigens are thus not simply eluted into the surrounding medium; their removal is an active process requiring, and linked to, temperature-sensitive processes in the cell. The loss of these antigens does not, however, involve the activity of the reactivated erythrocyte nucleus. If the ability of the erythrocyte nuclei to synthesize RNA is suppressed by massive irradiation of the erythrocytes with ultraviolet light prior to cell fusion, the hen-specific antigens

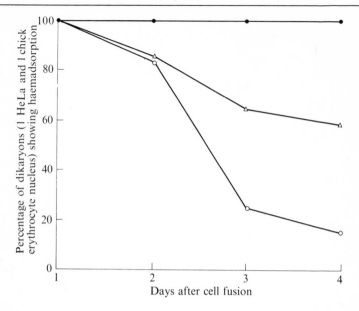

Fig. 3.4. The effect of temperature on the rate of loss of hen-specific antigens from the surface of heterokaryons made by fusing HeLa cells with chick embryo erythrocytes. The hen-specific antigens are eliminated at 37°C (○), more slowly at 28°C (△), and not at all at 20°C (●).

introduced into the heterokaryons are eliminated at much the same rate as in heterokaryons containing unirradiated erythrocyte nuclei. It thus appears that, at this stage, the activity of the erythrocyte nucleus does not determine the appearance of any new hen-specific antigen on the surface of the heterokaryon; nor does it influence the rate of disappearance of the antigens introduced during the process of cell fusion. On the other hand, the antigens synthesized by the recipient cells (human or mouse antigens as the case may be) do not disappear from the surface of the heterokaryons. It therefore seems likely that the disappearance of the hen-specific antigens reflects their gradual displacement by human or mouse antigens, which, one may suppose, are continually produced during the normal growth of the cell. If this is so, the rate of disappearance of the hen-specific antigens provides a direct measure of the turnover of species-specific antigens at the cell surface.

In one important respect the reactivation of the erythrocyte nuclei during the first 2 or 3 days after cell fusion is incomplete. Although these nuclei undergo great enlargement and resume the synthesis of RNA and DNA,[8-10] they do not develop normal nucleoli. On the third, and occasionally on the second, day after cell fusion, some erythrocyte nuclei develop small structures, which, under the light microscope, appear to be rudimentary nucleoli; but the prominent nucleoli characteristic of tissue cells in culture are not seen. Since the erythrocytes are taken from normal animals, there is no reason to suspect a genetic defect in this respect; and electron microscopy has indeed revealed structures in hen erythrocyte nuclei which, on morphological grounds, have been identified as the fibrillary component of the nucleolus.[11,12] One might therefore suppose either that the human or mouse cytoplasm is, in some unidentified way, an inadequate environment for the hen erythrocyte nucleus, or that these heterokaryons do not survive long enough as multinucleate cells to permit the nucleolus to develop fully in the erythrocyte nucleus. Within 4 days of cell fusion, the human or mouse nuclei in virtually all the heterokaryons enter mitosis. In some cases, this mitosis is irregular and results in the death of the cell. In others, the erythrocyte nucleus, which may not yet have completed the replication of its DNA, is destroyed by the process of 'chromosome pulverization'[13-16] when mitosis takes place; in others again, the erythrocyte nucleus appears to take part in the mitosis, so that nuclear fusion occurs.[17] In one way or another, mitosis thus results in the progressive disappearance from the cultures of heterokaryons in which a discrete erythrocyte nucleus can be recognized.

In order to permit development of the erythrocyte nucleus within the heterokaryon for a longer period, the recipient cells were therefore subjected to an appropriate dose of gamma radiation. The irradiated cells continued to grow for up to 3 weeks without undergoing mitosis and thus permitted the further development of the erythrocyte nucleus as a discrete entity. In these irradiated cells, nucleoli began to appear in the erythrocyte nuclei on the third day after cell fusion (Fig. 3.3) and became progressively larger. By the eleventh day, more than 80 per cent of the erythrocyte nuclei contained one or

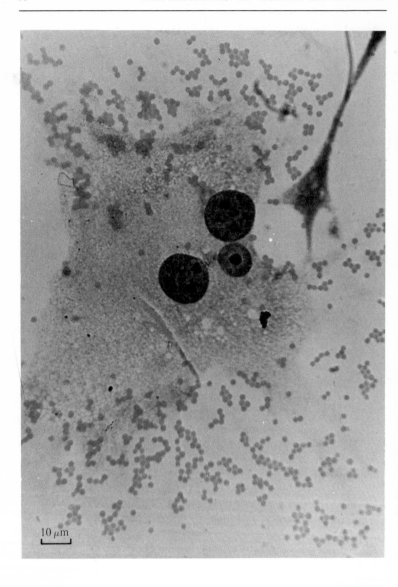

Fig. 3.5. A heterokaryon containing two irradiated mouse nuclei and one hen erythrocyte nucleus, 11 days after cell fusion. This is now a typical radiation giant cell. The reactivated erythrocyte nucleus has a prominent nucleolus, and hen-specific antigens have reappeared on the surface of the cell.

two readily identifiable nucleoli (Fig. 3.5). Hen-specific antigens could be detected on the surface of virtually all these
irradiated heterokaryons immediately after cell fusion and for
about 24 hours thereafter. These antigens were then progressively eliminated, but somewhat more slowly than in heterokaryons in which the recipient cell had not been irradiated.
By the sixth day, no hen-specific antigen could be detected in
any of the cultures. On the eighth day, however, traces of hen-
specific antigen began to reappear in some of the cells: the
antigen was first observed on the tips of elongated cytoplasmic
processes. On succeeding days, increasing numbers of cells

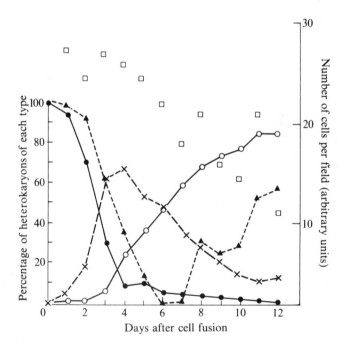

FIG. 3.6. Reappearance of hen-specific antigens on the surface of heterokaryons made
by fusing irradiated mouse cells with adult hen erythrocytes. The relationships between
the enlargement of the erythrocyte nucleus, the development of nucleoli within it, and
the disappearance and reappearance of the hen-specific surface antigens are shown.
□, total number of cells; ●, heterokaryons with unenlarged erythrocyte nuclei;
×, heterokaryons with enlarged erythrocyte nuclei, but no visible nucleoli; ○, heterokaryons with enlarged erythrocyte nuclei containing visible nucleoli; ▲, heterokaryons showing hen-specific surface antigens.

showed the presence of the antigen, which could now be detected all over the periphery of the cells. The amount of antigen per cell, as judged by the intensity of the haemadsorption reaction, continued to increase until, by the eleventh day, most of the heterokaryons showed strong haemadsorption over the whole of their periphery. The intensity of the haemadsorption greatly exceeded that seen in heterokaryons immediately after cell fusion. This sequence of events is illustrated in Figs. 3.2, 3.3, and 3.5 and plotted, as a function of time, in Fig. 3.6.

When the same experiment was done with erythrocytes from 12-day-old chick embryos, essentially similar results were obtained, except that the whole process took place more rapidly. Nucleoli began to appear in the erythrocyte nuclei on the second day after fusion, and the *de novo* appearance of chick-specific surface antigens took place before the antigens introduced during cell fusion were completely eliminated (Fig. 3.7). When erythrocytes from even younger chick embryos were used, the appearance of nucleoli and of new surface antigens occurred even sooner.[18] In all cases, there was a clear correlation between the speed with which the nucleoli developed and the time at which the species-specific antigens reappeared on the surface of the cells. These experiments thus indicated that hen erythrocyte nuclei, operating in the cytoplasm of cells from widely different animal species, were capable of determining the appearance of hen-specific surface antigens in these cells; but the reactivated erythrocyte nuclei did not determine the appearance of these antigens until they developed nucleoli.

Soluble enzyme

Since these species-specific antigens were detected only when they appeared at the surface of the cell, the possibility existed that there might be a lag between the time of their synthesis and the time of their detection by the haemadsorption technique. If this lag were considerable, the antigens might well have been synthesized before the appearance of nucleoli in the erythrocyte nuclei. In that case, the observed association between the appearance of the nucleoli and the appearance of the surface antigens might simply be fortuitous. It was therefore obviously necessary to examine the behaviour of other proteins which might be determined by the erythrocyte nu-

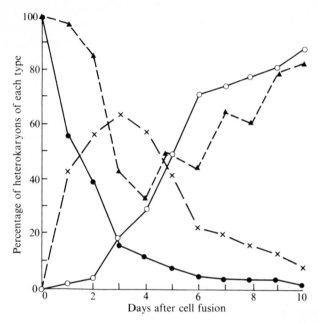

FIG. 3.7. Reappearance of hen-specific antigens on the surface of heterokaryons made by fusing irradiated mouse cells with chick embryo erythrocytes. The relationships between the enlargement of the erythrocyte nucleus, the development of nucleoli within it, and the disappearance and reappearance of the hen-specific surface antigens are shown. ●, heterokaryons with unenlarged erythrocyte nuclei; ×, heterokaryons with enlarged erythrocyte nuclei, but no visible nucleoli; ○, heterokaryons with enlarged erythrocyte nuclei containing visible nucleoli; ▲, heterokaryons showing hen-specific surface antigens.

cleus, and especially soluble proteins that did not form part of a larger structural organization. A soluble enzyme was the obvious choice, and the enzyme inosinic acid pyrophosphorylase (E.C.2.4.2.8.), which catalyses the condensation of hypoxanthine with ribosyl phosphate, was chosen for study.[19] The enzyme is essential for the incorporation of hypoxanthine into nucleic acids and may thus be assayed either directly in a cell homogenate[19,20] or indirectly in the intact cell by measuring the incorporation of labelled hypoxanthine.[21] This enzyme was chosen because a line of mouse cells was available which lacked inosinic acid pyrophosphorylase activity (A_9 cells).[22]

When A_9 cells are exposed to tritiated hypoxanthine and

then subjected to appropriate autoradiography, only a trivial amount of radioactivity is found to be incorporated into nucleic acids. When erythrocyte nuclei are introduced into irradiated A_9 cells, the heterokaryons also initially show very little incorporation of hypoxanthine. The enlargement and reactivation of the erythrocyte nuclei produce no change in this respect, until nucleoli make their appearance in the erythrocyte nuclei.

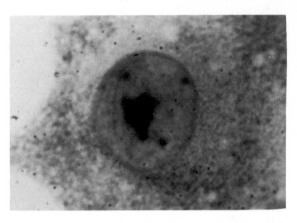

FIG. 3.8. Autoradiograph of an A_9 cell exposed for 4 h to tritiated hypoxanthine. There is very little incorporation of the label. Under such conditions a normal mouse cell would be very heavily labelled.

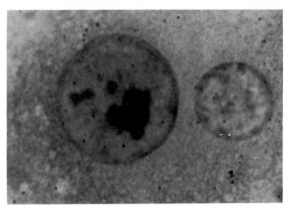

FIG. 3.9. Autoradiograph of an A_9-chick erythrocyte heterokaryon in which the erythrocyte nucleus has been reactivated but has not yet developed a nucleolus. The cell has been exposed for 4 h to tritiated hypoxanthine, but there is still very little incorporation of label.

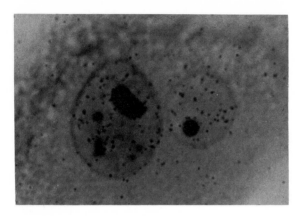

Fig. 3.10. Autoradiograph of an A_9-chick erythrocyte heterokaryon in which the erythrocyte nucleus shows early development of the nucleolus. The cell has been exposed for 4 h to tritiated hypoxanthine. Both the A_9 and the erythrocyte nucleus are now clearly labelled. The cell has acquired the ability to incorporate hypoxanthine into nucleic acid.

When this occurs, the ability of the heterokaryons to incorporate hypoxanthine shows a sharp increase, and autoradiographs begin to show RNA labelling in many of the cells. This sequence of events is illustrated in Figs. 3.8–3.10. On further cultivation, the ability of the heterokaryons to incorporate hypoxanthine and the number of erythrocyte nuclei showing nucleoli continue to rise *pari passu* (Fig. 3.11). At all times, those heterokaryons in which the erythrocyte nuclei have not yet developed nucleoli show no significant increase in hypoxanthine incorporation (Fig. 3.12). Increasing the number of erythrocyte nuclei per cell does not produce an earlier increase in hypoxanthine incorporation. Irrespective of the number of erythrocyte nuclei present, the ability of a heterokaryon to incorporate hypoxanthine remains negligible so long as the erythrocyte nuclei have not developed nucleoli (Fig. 3.13). Nucleoli commonly begin to appear in all the erythrocyte nuclei in a heterokaryon at about the same time, and, when this occurs, there is, once again, a sharp increase in the ability of the cell to incorporate hypoxanthine (Fig. 3.14). Direct assay of the enzyme in cell homogenates confirms the findings obtained in intact cells by autoradiographic procedures. Very little inosinic acid pyrophosphorylase activity is initially de-

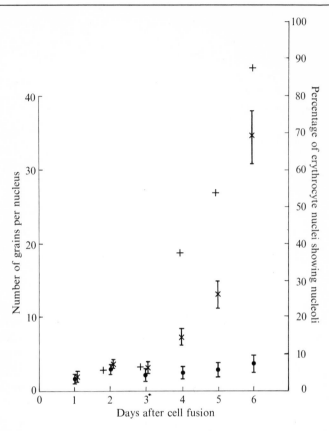

Fig. 3.11. The development of inosinic acid pyrophosphorylase activity in A_9-chick erythrocyte heterokaryons, as measured by their ability to incorporate tritiated hypoxanthine into nucleic acid. The incorporation of hypoxanthine in the heterokaryons is initially only marginally greater than that in A_9 cells alone; but when the erythrocyte nuclei develop nucleoli, this incorporation increases markedly. ●, A_9 cells; ×, heterokaryons; +, erythrocyte nuclei showing nucleoli.

tected in the heterokaryons, but, when nucleoli begin to appear in the erythrocyte nuclei, the enzyme activity rises sharply and continues to rise as the development of nucleoli proceeds (Fig. 3.15).

The fact that the soluble enzyme appears in the heterokaryons at the same time as the hen-specific surface antigen makes it very improbable that there can be any important lag, relative to the time scale on which these experiments are carried

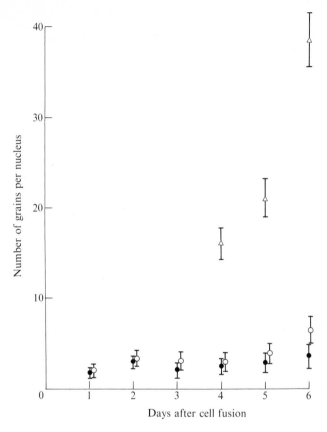

Fig. 3.12. A comparison between A_9–chick erythrocyte heterokaryons in which the erythrocyte nuclei have developed nucleoli and those in which they have not. The former show a marked increase in their ability to incorporate tritiated hypoxanthine; the latter are not much different from A_9 cells alone. ●, A_9 cells; △, erythrocyte nuclei showing nucleoli; ○, erythrocyte nuclei not showing nucleoli.

out, between the synthesis of the antigen and its appearance on the cell surface; and we are therefore faced with the conclusion that an enzyme and an antigen, which are neither structurally nor functionally related, both begin to be synthesized in the heterokaryon when nucleoli appear in the erythrocyte nuclei, and not before. The very fact that the antigen and the enzyme are so completely unrelated makes it unnecessary to entertain seriously the idea that the genes specifying these particular

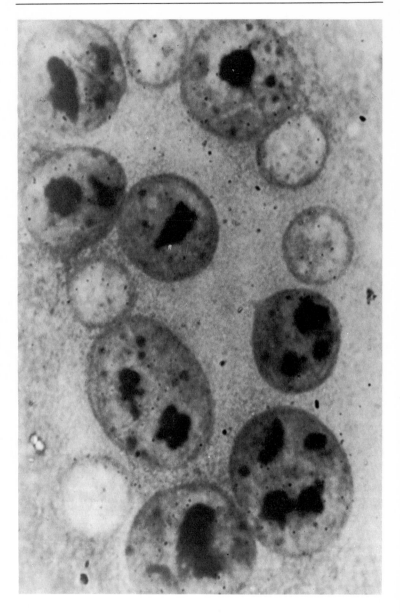

FIG. 3.13. A heterokaryon containing several A_9 and chick erythrocyte nuclei, exposed for 4 h to tritiated hypoxanthine. The erythrocyte nuclei have been reactivated but have not yet developed nucleoli. There is little incorporation of the label.

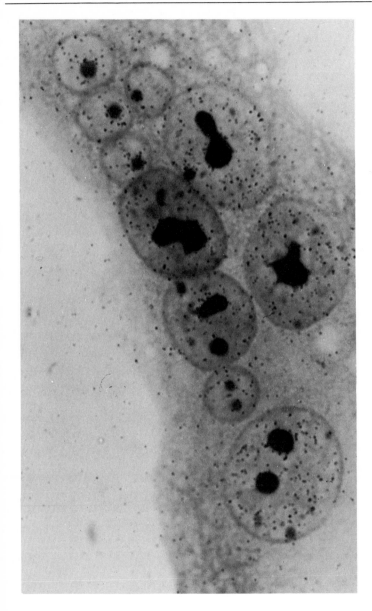

Fig. 3.14. A heterokaryon containing several A_9 and chick erythrocyte nuclei, exposed for 4 h to tritiated hypoxanthine. The erythrocyte nuclei have now developed nucleoli and all the nuclei are clearly labelled.

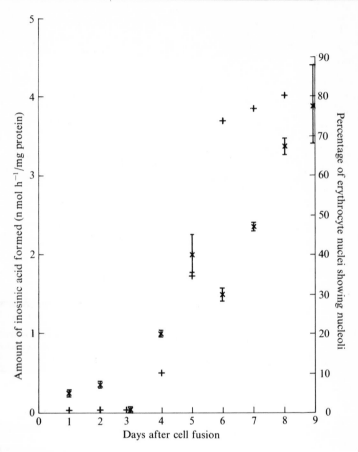

Fɪɢ. 3.15. The development of inosinic acid pyrophosphorylase activity in A₉–chick erythrocyte heterokaryons, as measured by direct assay of the cell homogenate. There is a marked increase in enzyme activity when the erythrocyte nuclei in the heterokaryon develop nucleoli. ×, inosinic acid pyrophosphorylase activity; +, erythrocyte nuclei showing nucleoli.

proteins begin to be transcribed only when a nucleolus appears in the erythrocyte nucleus. It is difficult to imagine why the transcription of just these genes should be delayed for some days and then begin simultaneously; nor are there any grounds for believing that these particular proteins are involved in nucleolar function. It seems much more reasonable to assume that the kinetics observed in the synthesis of these two unrelated

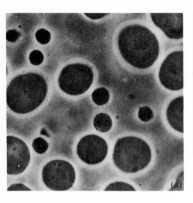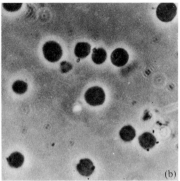

FIG. 3.16. (a) HeLa nuclei and erythrocyte nuclei isolated from heterokaryons 48 h after cell fusion. (b) Erythrocyte nuclei from the same preparation as (a), but separated from the HeLa nuclei by density gradient centrifugation.

proteins would also be observed with other proteins, and that the ability of the erythrocyte nucleus to determine the synthesis of *any* proteins would be conditional on the development of nucleolar activity. In any case, it was on this assumption that the investigation proceeded.

The RNA made in the erythrocyte nucleus

It was clearly of overriding importance to determine what kind of RNA was made in the erythrocyte nuclei before and after they developed nucleoli. A technique was therefore devised to permit re-isolation of the nuclei from the heterokaryons and separation of the reactivated erythrocyte nuclei from the human or mouse nuclei (Fig. 3.16).[3,23] The heterokaryons were exposed to radioactive RNA precursors at various times after cell fusion, and the RNA synthesized in the two sorts of nuclei was examined by sucrose gradient centrifugation. Three types of high molecular weight RNA are normally revealed in cell nuclei by sucrose gradient sedimentation: two components detected by ultraviolet absorption and having sedimentation coefficients in conventional sucrose gradients of approximately 28S and 16S; and a third detected by its radioactive content only and having a very polydisperse distribution. This 'polydisperse' RNA sediments largely as a broad peak ahead of the 28S component, but the radioactivity usually trails through the

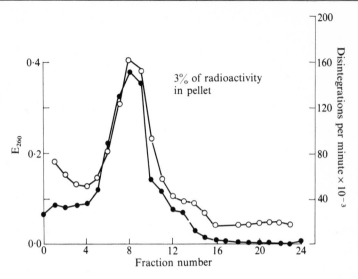

Fig. 3.17. Sucrose density gradient of the RNA extracted from HeLa nuclei isolated from HeLa–erythrocyte heterokaryons 48 h after cell fusion. The cells were exposed to a radioactive RNA precursor for 4 h before enucleation. The bulk of the radioactivity has been incorporated into the 28S and 16S RNA components. Very little radioactivity has been incorporated into 'polydisperse' RNA, which, under these conditions of centrifugation, sediments to the bottom of the centrifuge tube. ◯, E_{260}; ●, radioactivity.

28S and 16S components, thus making it difficult to decide how much label is actually in the 28S and 16S peaks and how much is simply coincident with them. It has been shown, however, that the sedimentation of the 'polydisperse' RNA is much more sensitive to changes in ionic strength and divalent cation concentration than the sedimentation of the 28S and 16S components;[24] and conditions can be found in which the 'polydisperse' RNA forms aggregates which sediment to the bottom of the centrifuge tube, while the 28S and 16S components remain in the body of the gradient.[24] In this way the 'polydisperse' RNA can be largely separated from the two major components. The RNA made by the two sets of nuclei in the heterokaryon was therefore examined under these conditions.

When heterokaryons in which the erythrocyte nuclei had not yet developed nucleoli were exposed for 4 hours to a radioactive RNA precursor, it was found that virtually all the radioactivity in the human or mouse nuclei had been incor-

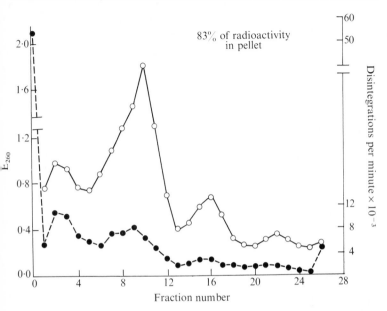

FIG. 3.18. Sucrose density gradient of the RNA extracted from the reactivated erythrocyte nuclei isolated from the same heterokaryons as those used to provide the HeLa nuclei analysed in Fig. 3.17. At this stage, the great majority of the erythrocyte nuclei had undergone enlargement, but very few of them had yet developed visible nucleoli. Most of the radioactivity has been incorporated into 'polydisperse' RNA, which, under these conditions, sediments to the bottom of the centrifuge tube. ○, E_{260}; ●, radioactivity.

porated into the 28S and 16S RNA components: only 3 per cent of the radioactivity was in 'polydisperse' RNA which sedimented to the bottom of the centrifuge tube (Fig. 3.17). But, in the erythrocyte nuclei, more than 80 per cent of the radioactivity was in 'polydisperse' components at the bottom of the tube, and very little label was in the 28S and 16S components (Fig. 3.18). Since this small incorporation of radioactivity into the 28S and 16S components could easily be accounted for by the few human or mouse nuclei that contaminated the preparations of erythrocyte nuclei, it seemed reasonable to conclude that the erythrocyte nuclei in which nucleoli had not yet developed synthesized 'polydisperse' RNA, but very little, if any, normal mature 28S and 16S RNA. When the erythrocyte nuclei were examined after they had developed nucleoli, the pattern of RNA labelling in them closely resembled that seen in the human or mouse nuclei in the same cells. Autoradiographic studies demonstrate that this 'polydisperse' RNA is synthesized, *grosso modo*, all over the nucleus and must therefore involve the activity of a very large number of genes.[9] We are therefore faced with the further paradox that this high molecular weight RNA, representing the activity of a very large number of genes, can be synthesized by the erythrocyte nuclei for several days without determining the synthesis of any specific proteins; but when the erythrocyte nuclei begin to make RNA of ribosomal type (28S and 16S RNA), proteins specified by these nuclei begin to be synthesized.

There appeared to be three possible explanations for these observations. The 'polydisperse' RNA made by the erythrocyte nuclei before development of nucleoli might not contain any RNA carrying instructions for protein synthesis. This explanation seemed unattractive, since it entailed the improbable consequence that the genes specifying particular proteins were not transcribed at all for several days and then, for some obscure reason, began to be transcribed simultaneously when the nucleolus made its appearance. Moreover, if the 'polydisperse' RNA did not at least include some RNA that carried instructions for protein synthesis, it was difficult to see where this RNA was to be found, since RNA having initially 'polydisperse' sedimentation was the only product that the great bulk of the genetic material appeared to synthesize. A second possibility was that the 'polydisperse' RNA did contain specifications for the synthesis of proteins, but that it was unable to programme the pre-existing ribosomes in the heterokaryons. Since the erythrocytes made no appreciable cytoplasmic contribution to these heterokaryons[7] (mature erythrocytes, in any case, contain virtually no ribosomes), all the ribosomes initially present in the heterokaryons were human or mouse ribosomes; and it was conceivable that the RNA made on the hen chromosomes might not be able to programme human or mouse ribosomes. The extensive work on disrupted cell systems did not encourage the belief that an interaction of such great specificity was a prerequisite for the synthesis of specific proteins. The third possibility was that the 'polydisperse' RNA did contain instructions for the synthesis of proteins, but that some nucleolar

activity was essential for the transport of this RNA to the cytoplasm of the cell. On this view, the RNA made before the
development of the nucleolus would fail to be transported to
the cytoplasm of the cell and would be eliminated within the
nucleus.

This last possibility was further investigated by means of
experiments in which a microbeam of ultraviolet light was
used to inactivate individual nuclei within the heterokaryon.[3,25]
If, before they developed nucleoli, the erythrocyte nuclei were
unable to transfer the RNA which they synthesized to the cytoplasm of the cell, then 'anucleolate' erythrocyte nuclei in
heterokaryons in which the human or mouse nuclei had been
inactivated by the microbeam would not be expected to contribute to cytoplasmic RNA labelling when the cells were
exposed to a radioactive RNA precursor; but these erythrocyte
nuclei would be expected to contribute to cytoplasmic RNA
labelling once they had developed nucleoli. Heterokaryons
containing a single mouse nucleus and up to four reactivated
erythrocyte nuclei were selected for study. The mouse nuclei
were inactivated by the microbeam, and the level of cytoplasmic RNA labelling in the irradiated cells was measured
after they had been exposed to a radioactive RNA precursor
for periods up to 6 hours. In the same cultures normal mononucleate mouse cells in which the nucleus had also been inactivated by the microbeam served as controls, since, even in
mononucleate cells, a low level of cytoplasmic RNA labelling
persists after the nucleus has been irradiated. The measurements were made at various times after cell fusion: before any
of the erythrocyte nuclei in the heterokaryons had developed
nucleoli; at a stage when some nucleoli were visible; and after
many days when almost all erythrocyte nuclei showed well-
developed nucleoli. During the period in which the erythrocyte nuclei had not yet developed nucleoli, the level of cytoplasmic labelling in heterokaryons in which the mouse nucleus
had been inactivated was indistinguishable from that found in
normal mononucleate mouse cells in which the nucleus had been
inactivated (Fig. 3.19). This was the case even in heterokaryons
that contained several reactivated erythrocyte nuclei which
collectively synthesized very large amounts of RNA as judged
by the intensity of nuclear labelling in autoradiographs. It

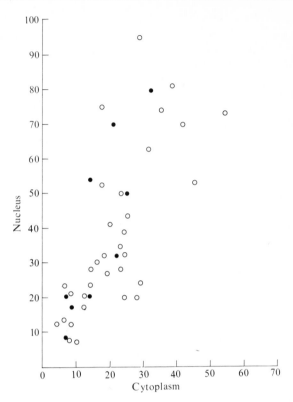

Fig. 3.19. Relationship between nuclear grain counts and cytoplasmic grain counts in heterokaryons and in single mouse cells. The heterokaryons contained one mouse nucleus and up to four chick erythrocyte nuclei which had been reactivated, but which had not yet developed nucleoli. The mouse nucleus was inactivated by ultraviolet light both in the heterokaryons and in the single mouse cells. The cells were exposed to a radioactive RNA precursor for 6 h. The ratio of nuclear to cytoplasmic RNA labelling in the heterokaryons in which the mouse nucleus had been inactivated was no different from that in the single mouse cells in which the nucleus had been inactivated. The reactivated erythrocyte nuclei, although they synthesize large amounts of RNA, do not make any detectable contribution to cytoplasmic RNA labelling at this stage. ○, heterokaryons; ●, single mouse cells.

thus appeared that, prior to the development of nucleoli, the reactivated erythrocyte nuclei, although they synthesized large amounts of RNA continuously for some days, did not transfer detectable amounts of this RNA to the cytoplasm of the cell. A similar experiment, done at a stage when some of the erythrocyte nuclei had developed nucleoli, showed that in some of the

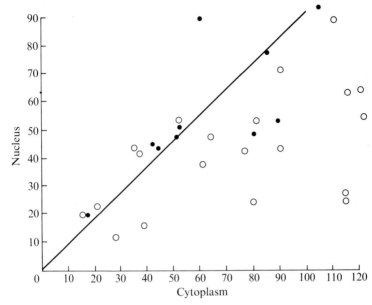

FIG. 3.20. A similar experiment to the one shown in Fig. 3.19, but done at an intermediate stage when some of the erythrocyte nuclei had begun to develop nucleoli. It will be seen that the cytoplasmic labelling is now greater in some of the heterokaryons than in the single mouse cells. The reactivated erythrocyte nuclei are beginning to contribute to cytoplasmic RNA labelling. ◯, heterokaryons; ●, single mouse cells.

heterokaryons the level of cytoplasmic labelling after irradiation of the mouse nuclei was now significantly greater than in mononucleate cells in which the nuclei had been irradiated (Fig. 3.20); and when the great majority of the erythrocyte nuclei had developed nucleoli, cytoplasmic labelling in the heterokaryons after irradiation of the mouse nuclei was at an altogether higher level than that found in the irradiated mononucleate cells (Fig. 3.21). Figures 3.22 and 3.23 show the levels of cytoplasmic RNA labelling in an unirradiated heterokaryon and in a heterokaryon in which the mouse nucleus was inactivated before the erythrocyte nucleus had developed a nucleolus. It will be seen that the erythrocyte nucleus continues to synthesize RNA but makes no contribution to cytoplasmic RNA labelling. Figure 3.24 shows, for comparison, a heterokaryon in which the mouse nucleus was inactivated after

G

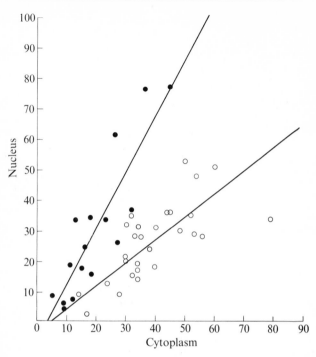

FIG. 3.21. A similar experiment to those shown in Figs. 3.19 and 3.20, but done at a stage when virtually all the erythrocyte nuclei had developed well-defined nucleoli. The cytoplasmic labelling in the heterokaryons is now decisively greater than in the single mouse cells. The erythrocyte nuclei are making a substantial contribution to cytoplasmic RNA labelling. ○, heterokaryons; ●, single mouse cells.

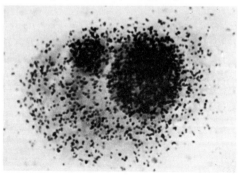

FIG. 3.22. Autoradiograph of a heterokaryon containing a mouse nucleus and a chick erythrocyte nucleus, exposed for 6 h to a tritiated RNA precursor. Both the mouse nucleus and the erythrocyte nucleus are very heavily labelled, and there is also sub-stantial cytoplasmic labelling.

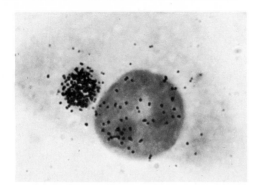

FIG. 3.23. Autoradiograph of another heterokaryon from the same preparation as the cell shown in Fig. 3.22. The mouse nucleus has been inactivated by ultraviolet light. The erythrocyte nucleus, which has not yet developed a nucleolus, is heavily labelled, but the cytoplasm contains very little radioactivity.

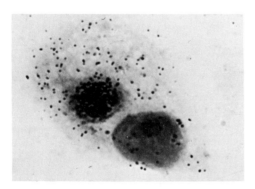

FIG. 3.24. Autoradiograph of a heterokaryon treated in the same way as the cell shown in Fig. 3.23, but after the erythrocyte nucleus had developed a nucleolus. The cytoplasm of the cell is now labelled.

the erythrocyte nucleus had developed a nucleolus. In this case, the erythrocyte nucleus does make a contribution to cytoplasmic RNA labelling. In the light of these experiments, the initial failure of the erythrocyte nuclei in the heterokaryons to determine the synthesis of specific proteins becomes understandable: it is not until these nuclei develop nucleoli that they acquire the ability to transmit genetic information to the cytoplasm of the cell.

The question, of course, at once arises whether this conclusion is restricted to erythrocyte nuclei reactivated under these extraordinary conditions, or whether the nucleolus plays a critical role in the transfer of genetic information in normal mononucleate cells. It was thought possible that some more direct evidence concerning the role of the nucleolus might be obtained by using the microbeam to irradiate the nucleolus alone.[25] Normal mononucleate HeLa cells were chosen for this study because many of these cells contained a single prominent nucleolus which lent itself admirably to selective irradiation. Nuclear and cytoplasmic RNA labelling were compared in four groups of cells from the one culture: (1) unirradiated cells; (2) cells with a single nucleolus which was selectively irradiated with a dose of ultraviolet light sufficient to reduce its ability to synthesize RNA to less than 5 per cent of normal; (3) cells in which a non-nucleolar part of the nucleus (nucleoplasm) was given the same dose of irradiation with the same microbeam; and (4) cells in which the whole nucleus was irradiated with a larger microbeam. The cells were again exposed to a radioactive RNA precursor for periods up to 6 hours after irradiation, and the pattern of RNA labelling analysed from autoradiographs. The results are shown in Figs. 3.25 and 3.26 and in Table 3.1. It will be seen that inactivation of the nucleolus alone reduces cytoplasmic RNA labelling by about 90 per cent, even though substantial RNA synthesis continues in the rest of the nucleus. That this is not a non-specific effect of the radiation is shown by the fact that comparable irradiation of the non-nucleolar part of the nucleus does not eliminate cytoplasmic RNA labelling. Irradiation of the whole nucleus, despite the much larger dose of ultraviolet light delivered to the cell, is hardly more effective in reducing cytoplasmic labelling than irradiation of the nucleolus alone (Fig. 3.25). It is clear that, in the absence of a functional nucleolus, the RNA synthesized in the rest of the nucleus is not transported in detectable amounts to the cytoplasm of the cell. These observations on the patterns of RNA labelling in mononucleate cells in which nucleolar activity is eliminated thus support the conclusions drawn from the experiments on erythrocyte nuclei reactivated in heterokaryons. The results in both cases indicate that the nucleolus is involved in the transfer, not only of the RNA made at the nucleolar site, but also of the RNA made elsewhere in the nucleus. Experimental error does not, of

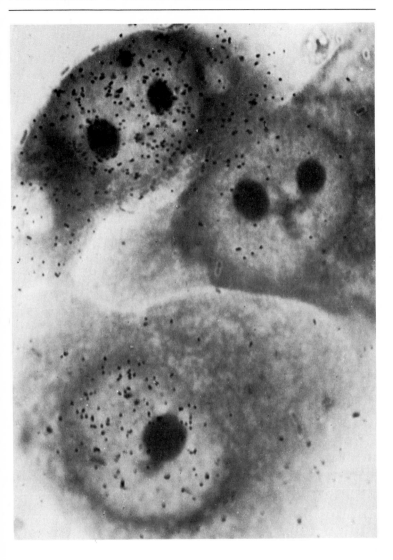

Fig. 3.25. Autoradiograph showing a normal HeLa cell (top left), a HeLa cell in which the whole nucleus has been irradiated with ultraviolet light (top right) and a HeLa cell in which the nucleolus only has been irradiated (bottom). The preparation was exposed to a radioactive RNA precursor for 6 h. Irradiation of the nucleolus alone, despite continued synthesis of RNA elsewhere in the nucleus, reduces cytoplasmic labelling to a level comparable with that seen in the cell in which the whole nucleus has been irradiated.

course, permit the statement that all the high molecular weight RNA made in the nucleus is dependent on nucleolar activity for its transport to the cytoplasm; but it does appear that the great bulk of the RNA made on the chromosomes fails to be transferred to the cytoplasm of the cell if the nucleolus is artificially inactivated or if it has not yet developed. It therefore seems very improbable that the RNA made in the nucleolus and the RNA made elsewhere in the nucleus can be transferred to the cytoplasm independently. All the results obtained in the present study indicate that the transport of both these classes of RNA is co-ordinated by a single process which is located at or near the nucleolus.

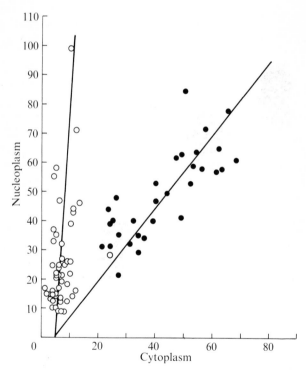

Fɪɢ. 3.26. Comparison of the number of grains over the cytoplasm with the number of grains over the non-nucleolar parts of the nucleus (nucleoplasm) in unirradiated HeLa cells (●) and in HeLa cells in which the nucleolus only was irradiated (○). The cells were exposed to a radioactive RNA precursor for 6 h. Irradiation of the nucleolus alone abolishes cytoplasmic RNA labelling, despite the fact that RNA synthesis continues in the rest of the nucleus.

TABLE 3.1

Grain counts over nucleoplasm and cytoplasm in HeLa cells in which either the whole nucleus or the nucleolus was irradiated

	No. cells	Total no. grains		Mean counts nucleo-plasm	Mean counts cyto-plasm
		Nucleo-plasm	Cyto-plasm		
Unirradiated cells	37	2016	1885	54·5	50·9
Nucleolus irradiated	50	1284	347	25·7	6·9
Whole nucleus irradiated	43	207	217	4·8	5·0
Nucleoplasm irradiated	24	888	806	37·0	33·6

These findings thus argue against any model for the transfer of information from nucleus to cytoplasm that postulates that the RNA carrying the information for protein synthesis diffuses from the nucleus into the cytoplasm and there attaches to pre-existing ribosomes; and they also argue against any model that postulates the passage of cytoplasmic ribosomes into the nucleus in order to release this RNA from the genes or serve as a vehicle for its transport to the cytoplasm. The hen erythrocyte nucleus in the heterokaryon does not begin to transfer detectable amounts of RNA to the cytoplasm of the cell and does not initiate the synthesis of hen-specific proteins until it develops its own nucleolus and begins to synthesize its own 28S RNA (and presumably its own ribosomes). The pre-existing ribosomes in the cytoplasm of the hybrid cell cannot accomplish the transfer of genetic information from the 'anucleolate' erythrocyte nucleus.

This fact has important consequences for our understanding of the mechanisms by which genetic information is expressed. The flow of genetic information from the nucleus to the cytoplasm of the cell appears to be coupled to the flow of ribosomes; and the flow of ribosomes appears in turn to be directly dependent upon the synthetic activities of the nucleolus. The nucleolus may thus be envisaged as the centre of some general regulatory mechanism that governs the flow, not only of structural RNA components, whatever these might be, but also of those components in the ribosomal complex that carry the specifications for protein synthesis. If, as all the present experiments suggest, there is no independent flow of information to the cytoplasm of the cell, if the structural components of the ribosomal complex and those carrying the specifications for protein synthesis pass to the cytoplasm together, it follows that, under steady-state conditions, the RNA that carries the specifications for protein synthesis must have the same overall stability in the cytoplasm as the other RNA components in the ribosomal complex. To seek to identify the RNA that carries the specifications for protein synthesis by its relative instability will, on this view, prove to be a search for a will o' the wisp. Moreover, if genetic information cannot pass to the cytoplasm without a concomitant flow of ribosomes, 're-programming' the cytoplasm of a cell must entail the replacement of one family of 'programmed' ribosomes by another. When cells are called upon to do something radically new, one would then expect to find not only an increase in the flow of ribosomal RNA from the nucleus, but also a breakdown of the ribosomal RNA already in the cytoplasm. And this is precisely what one does find.[26]

This is hardly an appropriate point at which to begin a speculative discussion about how, in chemical terms, the nucleolus might exercise this general control over the flow of genetic information. I have already committed this sin elsewhere.[3,27] But I hope I may be forgiven for introducing, as a last word, one of my oldest hobby-horses. My entry into the field of nucleocytoplasmic relationships began in 1959, when J. W. Watts and I stumbled upon the phenomenon of intranuclear RNA turnover.[28,29] The work of a decade in my own laboratory[30-36] and many others[37-44] has not yet provided what I would regard as a satisfactory explanation for this phenomenon. I think the experiments described in this chapter may do so. If, in order to be transported to the cytoplasm of the cell, the RNA made on the chromosomes must first be 'engaged' by some mechanism that depends on nucleolar activity, the limiting step in the process of information transfer will not be the synthesis of the RNA that carries the genetic information, but the 'engagement' of this RNA by the nucleolar mechanism. And if the nucleolus is to play a regulatory role, which it clearly does, the 'engagement' mechanism must be sensitive to environmental influences, so that, as conditions change, a varying amount of the RNA made on the chromosomes fails to be 'engaged'. I should like to suggest that intranuclear RNA turn-over is simply the intranuclear elimination of the RNA that fails to be 'engaged'. If this is so, one would expect intranuclear turnover of RNA to be more pronounced in nuclei that have a poorly developed nucleolar system or nuclei in which the nucleolar system has been impaired. The study of a wide range of experimental material confirms that this is indeed the case.[3,25,28,29,38,40,45] But there is a difficulty. Although RNA having 'polydisperse' sedimentation properties appears to be the only primary product of the vast majority of genes, no RNA having the physical properties of this 'polydisperse' RNA can be detected in the cytoplasm of the cell. We must therefore assume, since information from at least some of these genes does reach the cytoplasm, that the 'polydisperse' RNA, during the process of 'engagement' and transfer to the cyto-plasm, undergoes some secondary structural modifications which endow it with different sedimentation properties. But then, how are we to identify it in the cytoplasm of the cell? No one, at this moment, is at all sure. Which RNA in the ribo-somal complex is the RNA that carries the specifications for protein synthesis? That is the question.

REFERENCES

1. HARRIS, H. (1969). The expression of genetic information. A study with hybrid animal cells. *Ciba Symposium* on *Control processes in multicellular organisms*. In press.
2. RILEY, M., PARDEE, A. B., JACOB, F., and MONOD, J. (1960). On the expression of a structural gene. *J. molec. Biol.* **2**, 216.
3. HARRIS, H., SIDEBOTTOM, E., GRACE, D. M., and BRAMWELL, M. E. (1969). The expression of genetic information: a study with hybrid animal cells. *J. Cell Sci.* **4**, 499.
4. WATKINS, J. F. and GRACE, D. M. (1967). Studies on the surface antigens of interspecific mammalian cell heterokaryons. *J. Cell Sci.* **2**, 193.
5. ESPMARK, J. H. and FAGRAEUS, A. (1965). Identification of the species of origin of cells by mixed haemadsorption: a mixed antiglobulin reaction applied to monolayer cell cultures. *J. Immun.* **94**, 530.

H

6. COOMBS, R. R. A., MARKS, J., and BEDFORD, D. (1956). Specific mixed agglutination: mixed erythrocyte-platelet antiglobulin reaction for the detection of platelet antibodies. *Br. J. Haemat.* **2**, 84.

7. SCHNEEBERGER, E. E. and HARRIS, H. (1966). An ultrastructural study of interspecific cell fusion induced by inactivated Sendai virus. *J. Cell Sci.* **1**, 401.

8. HARRIS, H. (1965). Behaviour of differentiated nuclei in heterokaryons of animal cells from different species. *Nature, Lond.* **206**, 583.

9. HARRIS, H. (1967). The reactivation of the red cell nucleus. *J. Cell Sci.* **2**, 23.

10. BOLUND, L., Ringertz, N. R., and HARRIS, H. (1969). Changes in the cytochemical properties of erythrocyte nuclei reactivated by cell fusion. *J. Cell Sci.* **4**, 71.

11. TOOZE, J. and DAVIES, H. G. (1967). Light- and electron-microscope studies on the spleen of the newt *Triturus cristatus*: the fine structure of erythropoietic cells. *J. Cell Sci.* **2**, 617.

12. SMALL, J. V. and DAVIES, H. G. Personal communication.

13. SAKSELA, E., AULA, P., and CANTELL, K. (1965). Chromosomal damage of human cells induced by Sendai virus. *Annls Med. exp. Biol. Fenn.* **43**, 132.

14. NICHOLS, W. W., AULA, P., LEVAN, A., HENEEN, W., and NORRBY, E. (1967). Radioautography with tritiated thymidine in measles- and Sendai virus-induced chromosome pulverization. *J. Cell Biol.* **35**, 257.

15. KATO, H. and SANDBERG, A. A. (1968). Chromosome pulverization in Chinese hamster cells induced by Sendai virus. *J. natn. Cancer Inst.* **41**, 1117.

16. KATO, H. and SANDBERG, A. A. (1968). Cellular phase of chromosome pulverization induced by Sendai virus. *J. natn. Cancer Inst.* **41**, 1125.

17. HARRIS, H., WATKINS, J. F., FORD, C. E., and SCHOEFL, G. I. (1966). Artificial heterokaryons of animal cells from different species. *J. Cell Sci.* **1**, 1.

18. COOK, P. R. Unpublished results.

19. HARRIS, H. and COOK, P. R. (1969). Synthesis of an enzyme determined by an erythrocyte nucleus in a hybrid cell. *J. Cell Sci.* **5**, 121.

20. LITTLEFIELD, J. W. (1963). The inosinic acid pyrophosphorylase activity of mouse fibroblasts partially resistant to 8-azaguanine. *Proc. natn. Acad. Sci. U.S.A.* **50**, 568.

21. SUBAK-SHARPE, H. (1965). Biochemically marked variants of the Syrian hamster fibroblast cell line BHK21 and its derivatives. *Expl Cell Res.* **38**, 106.

22. LITTLEFIELD, J. W. (1964). Three degrees of guanylic acid-inosinic acid pyrophosphorylase deficiency in mouse fibroblasts. *Nature, Lond.* **203**, 1142.

23. FISHER, H. W. and HARRIS, H. (1962). The isolation of nuclei from animal cells in culture. *Proc. R. Soc.* B**156**, 521.

24. BRAMWELL, M. E. and HARRIS, H. (1967). The origin of the polydispersity in sedimentation patterns of rapidly labelled nuclear ribonucleic acid. *Biochem. J.* **103**, 816.

25. SIDEBOTTOM, E. and HARRIS, H. (1969). The role of the nucleolus in the transfer of RNA from nucleus to cytoplasm. *J. Cell Sci.* In press.
26. TATA, J. R. (1969). Hormonal control of metamorphosis and morphogenesis. *Ciba Symposium* on *Control processes in multicellular organisms.* In press.
27. HARRIS, H. (1968). *Nucleus and cytoplasm.* Clarendon Press, Oxford.
28. WATTS, J. W. and HARRIS, H. (1959). Turnover of nucleic acids in a non-multiplying animal cell. *Biochem. J.* **72**, 147.
29. HARRIS, H. (1959). Turnover of nuclear and cytoplasmic ribonucleic acid in two types of animal cell, with some further observations on the nucleolus. *Biochem. J.* **73**, 362.
30. HARRIS, H. and WATTS, J. W. (1962). The relationship between nuclear and cytoplasmic ribonucleic acid. *Proc. R. Soc.* B**156**, 109.
31. HARRIS, H., FISHER, H. W., RODGERS, A., SPENCER, T., and WATTS, J. W. (1963). An examination of the ribonucleic acids in the HeLa cell with special reference to current theory about the transfer of information from nucleus to cytoplasm. *Proc. R. Soc.* B**157**, 177.
32. HARRIS, H. (1963). The breakdown of RNA in the cell nucleus. *Proc. R. Soc.* B**158**, 79.
33. HARRIS, H. (1963). Nuclear ribonucleic acid. *Prog. nucl. Acid Res.* **2**, 20. Academic Press, New York.
34. HARRIS, H. (1964). Breakdown of nuclear ribonucleic acid in the presence of actinomycin D. *Nature, Lond.* **202**, 1301.
35. WATTS, J. W. (1964). Turnover of nucleic acids in a multiplying animal cell. 2. Retention studies. *Biochem. J.* **93**, 306.
36. HARRIS, H. (1965). The short-lived RNA in the cell nucleus and its possible role in evolution. *Evolving genes and proteins* (eds. V. Bryson and H. J. Vogel), p. 469. Academic Press, New York.
37. EDSTRÖM, J-E. (1965). Chromosomal RNA and other nuclear RNA fractions. *Role of the chromosomes in development* (ed. M. Locke), p. 137. Academic Press, New York.
38. BRUNS, G. P., FISCHER, S., and LOWY, B. A. (1965). A study of the synthesis and interrelationships of ribonucleic acids in duck erythrocytes. *Biochem. biophys. Acta* **95**, 280.
39. SOEIRO, R., BIRNBOIM, H. C., and DARNELL, J. E. (1966). Rapidly labelled HeLa cell nuclear RNA. *J. molec. Biol.* **19**, 362.
40. ATTARDI, G., PARNAS, H., HWANG, M., and ATTARDI, B. (1966). Giant-size rapidly labelled nuclear RNA and cytoplasmic messenger RNA in immature duck erythrocytes. *J. molec. Biol.* **20**, 145.
41. HOUSSAIS, J.-F. and ATTARDI, G. (1966). High molecular weight non-ribosomal-type nuclear RNA and cytoplasmic messenger RNA in HeLa cells. *Proc. natn. Acad. Sci. U.S.A.* **56**, 616.
42. LAZARUS, H. M. and SPORN, M. B. (1967). Purification and properties of a nuclear exoribonuclease from Ehrlich ascites tumour cells. *Proc. natn. Acad. Sci. U.S.A.* **57**, 1386.
43. KIJIMA. S. and WILT, F. H. (1969). Rate of nuclear ribonucleic acid turnover in sea urchin embryos. *J. molec. Biol.* **40**, 235.
44. ARONSON, A. I. and WILT, F. H. (1969). Properties of nuclear RNA in

sea urchin embryos. *Proc. natn. Acad. Sci. U.S.A.* **62**, 186.

45. PERRY, R. P. (1963). Selective effects of actinomycin D on the intracellular distribution of RNA synthesis in tissue culture cells. *Expl Cell Res.* **29**, 400.

Author Index

Subject Index